ANNA-TERESA TYMIENIECKA
The World Phenomenology Institute

LOGOS AND LIFE:

THE PASSIONS OF THE SOUL AND THE ELEMENTS IN THE ONTO-POIESIS OF CULTURE

The Life-Significance of Literature

Published under the auspices of
The World Institute for Advanced Phenomenological Research and Learning
A-T. Tymieniecka, President

KLUWER ACADEMIC PUBLISHERS
DORDRECHT / BOSTON / LONDON

This monograph essay is the Tractatus Brevis in Volume XXVIII
in the Analecta Husserliana: Tymieniecka, A-T. (ed.),
Poetics of the Elements in the Human Condition. Part 3:
The Elemental Passions of the Soul. 1990
664 pp. cloth. ISBN 0-7923-0180-3

Published by Kluwer Academic Publishers,
P.O. Box 17, 3300 AA Dordrecht, The Netherlands.

Kluwer Academic Publishers incorporates the publishing programmes of
D. Reidel, Martinus Nijhoff, Dr W. Junk and MTP Press.

Printed in the Netherlands

ANNA-TERESA TYMIENIECKA

TRACTATUS BREVIS

ANNA-TERESA TYMIENIECKA

LOGOS AND LIFE
BOOK 3

THE PASSIONS OF THE SOUL AND THE ELEMENTS IN THE ONTO-POIESIS OF CULTURE

The Life-Significance of Literature

TABLE OF CONTENTS

SECTION TWO
THE PASSIONS AND THE ELEMENTS IN THE HUMAN CONDITION: LIGHT AND THE SEA

THE PRIMEVAL ELEMENTS AND THE ELEMENTAL PASSIONS OF THE SOUL

INTRODUCTION

THE CONTEXT OF THE NOVEL METAPHYSICAL VISION OF THE ONTO-POIESIS OF BEINGNESS

Traditional ontology and metaphysics inquired into the last principles and innermost structures of what is. They assumed the *status quo* of the world, the cosmos, and of objectivity as such. Overwhelmed by the power of the physical elemental forces with Nature, by then unreachable stars and planets and by their seeming endurance while human life surges, develops, and unfailingly declines toward extinction, the traditional ontologies sought in the notion "Being" an ever-present, and everywhere the same last principle that maintains whatever there is whether in stasis or in flux.

Modern philosophers shifted the quest for the last principles of what-there-is to the constitutive-envisioning power of the human mind. The human mind was thought to provide the explanation of how what-there-is — as it is — emerges within the transcendental circuit of human functioning.

With the surging of a new metaphysical vision which presides over the present inquiry, one in which *what-there-is* is seen to consist in an ever-renewed constructive progress, the emphasis shifts from "what-there-is" to the ways and modes of becoming. One further step taken beyond the metaphysical views of Whitehead, Bergson, etc., and we reach the real crux of the matter; *the vision of life.* It is, in fact, life which carries the flux of becoming. Becoming is not a haphazard topsy-turvy coming together and going asunder. It is the poiesis of life as a constructive progress which establishes the relative stability of instants of what-there-is. No longer can the notion of being function as a principle of the principles which sustain what-there-is. The principle, rather, is "beingness," which is what individualizes something and through which, as through a vehicle, life expands. This means that the

A-T. Tymieniecka (ed.), Analecta Husserliana, Vol. XXVIII, 9—12.

new vision of what-there-is-and-was, as it originates and unfurls in its beingness, which I have presented in my phenomenology of life, focuses precisely upon the life-course of this beingness in its generic progress.

Living beingness, the vehicle of life, carries the constructive progress of life in a twofold perspective. (1) It individualizes life's very own unfolding through the successive steps of its inward/outward functioning and introduces the concrete and limited to itself and those other living beings who partake of this segment of life's advance; it introduces vectors of internalization/externalization, of succession and duration, of advance and regress. This means an order already intrinsic to the circumambient world it simultaneously projects. But the great question emerges: What are the primogenital constructive coordinates of the order of the becoming? I have answered this question previously by stating that the life-world, the otherness, and the Human Condition are the three constructive coordinates of life's progressive advance in complexity.

(2) There is still the perspective of the evolutive progress of types in which the advance of individual living beingnesses proceeds. Indeed, evolving stepwise in advancing complexity, the types reach the phase of the Human Condition, zenith of the advance as far as we know it. The Human Condition is a set of functional virtualities which accounts for the emergence of the specific type of living beingness that is man within which the interiorization/exteriorization vectors of life assume a unique amplitude and significance.

This allows the human being to survey from within the without and to raise the final metaphysical question: What are the ultimate onto-poietic factors upon which are suspended the constructive coordinates within which the individualizing progress of life proceeds; and what are the first elemental factors which preside over the establishment and subsistence of life?

In the present study which is conducted within a metaphysical vision of life, I will attempt to investigate two of these onto-poietic factors in the great system of life: the element of light, which is present at the origin of elementary life and omnipresent within it; and the sea, an element which offers us the entire panorama of the human-life experience.

Both will reveal to us the specific elemental passions of the soul and their proficiencies for creating the specifically human significance of life. The fundamental perspective of this proto-metaphysical inquiry makes us aware of an enlightening truth: it is in the aesthetic engagement of the human being in the arts, but particularly in the fabulating creativity of literature, that the specifically human life-significance originates and unfolds. By seeking to unravel the originary sources and motives of the unfolding of culture — the specific outgrowth of the onto-poiesis of the human being as he endows his beingness with its essential features, tendencies, moves, aspirations — literature offers an incomparable access to the realm of our query. In reverse, these originary sources and motives — being hidden in the deepest recesses of our inner/outer functioning where there is determined the route of the unfolding of our human beingness through all the phases of life — have to be brought to light if we want to understand what aesthetics, art, and literature are about.

As we first absorb it, and on the surface, art, but especially literature as it flourishes in a given historical period, represents the gist of the culture of its period. As we know, the majority of writers follow the currents of thought of their time; they choose their words according to the presumed taste of a prospective reader. But this tandem relationship involves only the shallow surface of things.

At the onto-poietic depths where a work of art is generated, this relativism explains nothing. The progress of living beingness with its inward/outward orientation cannot be reductionistically explained by the conditions of its external situations: while there is reciprocal conditioning between the "inner" and the "outer" and life progress in all the phases of its unfolding is the result of this play which offers a wide range of variable possibilities, these varying possibilities for life's unfolding are not haphazard, or random; their principles are constant. These constants in life's progress, its inward/outward orientation, its dominant/submissive dimensions belong to the primeval laws of life. They are present wherever human life manifests itself, as well as in every phase of culture.

It is these constants of the onto-poiesis of life that we are going to investigate as we take the bull by its horns and raise the question: What is the vocation of literature?

In addition to their perennial philosophical pertinence, the insights to be gained into the variables and constants of human life and cultural existence could not now be more needed. In our times when unforeseeable marvels produced by human genius bedazzle man to the point of disorienting him so that instead of experiencing due enthusiasm he falls into bewilderment, doubt, and foreboding, it is of paramount significance, first, to rediscover the principles of life that are the best guide for the human being when he takes the tiller in his hand, and, second, to reawaken in the contemporary writer and critic, now half-lost in a maze of inspirations coming from immature theories and societal trends and given to engaging in spurious exercises displaying his virtuosity in new writing techniques and word games, the crucial role in culture and for the whole of humanity that is his.

SECTION I
LITERATURE IN ITS CREATION AND RECEPTION

CHAPTER ONE

THE VOCATION OF LITERATURE AND THE SUBLIMINAL FORGE OF THE LIFE SIGNIFICANCE OF LITERATURE

(a) *Does Literature Have Its Own Vocation?*

Does literature have a vocation? To ask this question means to ask whether it has a unique task to perform for the sake of human life and culture. This question is not identical to that of "Why does a writer engage in literary pursuits?" and yet these two questions overlap. There might be many circumstantial reasons for taking up literature, but to train oneself to be a writer means already that there is a most profound personal commitment to fulfill one's intimate wishes, ideals, and a task that looms on the far-off horizon, and this is the most fundamental reason. This amounts to saying that there is an innermost motivation for the writer's urge to write, to communicate something uniquely his own to a public, to the society of his time, and to enrich by his message — or even transform — the culture of his period or of all time even. It is between this urge and its fulfillment that we may situate the question of the significance which such an undertaking may, or should, or can have for those who will receive it.

The question of the vocation of literature means to ask whether there is a specific role which the exemplary — in Kant's sense — great works of literature, or even literature as *belles lettres*, is called upon to fulfill for the nurturing of the individual human being as well as of culture at large.

It has often been pointed out that the writer is a witness to his times; we may enlarge on this and propose that a great writer is a witness to the whole of humanity. Not only does he explore the forces and the struggles of his time, but he seeks to reveal their hidden reasons in human nature: he will ponder the reasons for the course of national and human destiny.

A-T. Tymieniecka (ed.), Analecta Husserliana, Vol. XXVIII, 13—20.

These questions have been so much debated, and yet it is always useful to review them. What is of primary interest, however, in our bringing forth at this point these debates on the role of literature is the fact that in defining the role of literature we encroach upon philosophy. To reveal reasons is, in fact, the main task of the philosophical quest.

The question of the relationship between literature and philosophy has never been so pressing as at the present. From the dawn of the Occidental culture there has been a pervading intertwining of literature and philosophy that is obvious in all of the great works of literature as well as in certain strains of philosophy.

Great writers have never stopped at the description of life, at showing its intricacies, psychological underpinnings, and conflicts; they venture further attempting to fathom the elevation or degradation of the human spirit, groping to elucidate it. Dostoyevsky in all of his novels shows us the bottomless yearning of the human being for something higher than his narrow world, a yearning that cannot be satisfied by anything he undertakes and which remains looming over life like a melancholy doom awaiting when all of life's forces and life itself is spent. Chekhov's short stories and plays literally swim in a metaphysical climate of wonder over human fate, a climate which is the quintessence of their sense. The greatness of a literary work consists precisely in the fact that an author like Manzoni can, as in his *The Betrothed*, inspire the reader with a vision of a marvelous harmony within human life which prevails over unforeseen turmoil, cataclysm, ignorance, and human misdeed. Balzac, who focuses so on the mores of his period, does not fail to give us profound insight into ineffable recesses of the human soul.

If these authors, to which a great many others could be added, offer an "intellectual" view of life, the world, human destiny, etc., this does not mean that we may, following Julian Marias, see it as the role of literature to offer us precisely that. As a matter of fact, in discussing the literary expression of philosophical thoughts that we have in Miguel de Unamuno's work, Marias compares the tasks of both philosophy and literature seeing their meeting point in the attempt *to present the world.* They diverge, however, in intent and modality of presentation; literature marshals cultural surroundings, concrete human lives, feelings, and experiences, as well as societal entanglements to show us the world, whereas philosophy, through the sole means of the intellect, seeks the rational ordering, reasons, and principles operating in the world. But what do we find when we enter into Unamuno's own work and into his own views on the relationship between the two approaches? As Marias himself emphasizes, each of Unamuno's novels is an autobiographical novel in the sense that each of their heroes present experiences lived by the author himself. Unamuno tells us, in fact, in his essay, "How a Novel Is Made," that "each novel, every work of fiction, every poem created by an author is, insofar as it is alive, autobiographical. . . . Each poetic personage created by the author embodies part of the author himself."[1] If this presentation of the innermost individual struggles of "a man in his body and bones," which Unamuno attempts, encroaches upon the territory of philosophy, this is true inasmuch as Unamuno situates the core of this autobiographical account on a level where the individual in question meets with all humanity; it is his conviction that the most essential striving of the human being is his striving to perdure beyond the limited moment, beyond the time which threatens to devour our life, by attempting to endow it with an innermost sense that makes it a "legend", "a dream of eternity"[2] (or that envy be the most insidious factor of our life-conduct). These are elements of a philosophical conception of the human being. However, even as he proclaims and defends these convictions revealed in his novels to be the absolutely valid truth about the human being and about life, he distances himself from philosophizing by denying the validity of philosophy. Although he extends the creative activity of the novelist presenting his own dream — the dream of his life — to the activity of the philosopher, he nevertheless departs from his dream as he then puts all his efforts into systematizing his insights. For, as Unamuno writes: "The system destroys the essence of a dream and with it the essence of life."[3]

Discussing the relation of literature to philosophy, Italo Calvino dismisses both the "intellectual" and the "existential" novel as passé. A short article by him on philosophy and literature begins with this statement:

> Philosophy and literature are embattled adversaries. The eyes of philosophers see through the opaqueness of the world, eliminate the flesh of it, reduce the variety of existing things to a spider's web of relationships between general ideas, and fix the rules according to which a finite number pawns moving on a chessboard exhausts a number of combinations that may even be infinite. Along come the writers, and they replace the abstract chessman with kings and queens, knights and castles, all with a name, a particular shape, and a series of attributes royal, equine, or ecclesiastic; instead of a chessboard, they roll out great dusty battlefields or stormy seas. So at this point, the

rules of the game are turned topsy-turvy, revealing an order quite different from that of the philosophers.[4]

But Calvino goes on, saying that those who discover this change in the rules are again philosophers, who then start arguing about the validity of their role in reaching the common goal of the "conquest of truth."

Philosophy and literature are caught in a constant contest as each attempts to absorb each other's task. Discussing the intellectual category of writers (e.g., Pirandello's drama, Huxley's *Brave New World*, Thomas Mann's *The Magic Mountain*, etc.), Calvino opines that their "intellectuality" is "the thing of the past; today an author would express their ideas in a scholarly study." In contrast, existential philosophers who in their novels clothe their ideas "fully in human flesh" do not "acquire a literary stringency of their own" (Sartre); the writer has to write "before the philosopher in order to interpret him."[5]

In his dissection of the various ways in which philosophy and literature invade each other's territories (e.g., in ethics) as they both attempt to interpret reality, Calvino emphasizes that various forms of contemporary fiction writing pursues types of philosophical inquiry. The French Tel Quel group for example, explores the games of language and experience, of fiction and reality, of language as such, and goes so far as to plumb the ontology of language, thus turning fiction writing into an austere speculative exercise.

This situation makes the question of the relation between literature and philosophy particularly timely and acute. In Calvino's own novels which are characterized by an intense metaphysically allegoric climate in which the everyday contours of human existence are dissolved, some quintessential aspects of this relationship appear; and as a penetrating literary critic to boot, he attempts his own analysis of their relationship. First, Calvino wants to show us that both philosophy and literature are at war "with each side confident of having taken a step ahead in the conquest of truth, or at least of *a* truth." Yet he emphasizes that both are aware of their using for their constructions the same raw material, which is words. The use, however, of words seems to draw a hazy line of demarcation — as he sees it in the case of phenomenological and existential philosophy — between the two, a frontier which ultimately cannot be crossed without (and he singles out existentialist literature, e.g., Sartre, here) one or the other falling by the wayside. Since the interpretation of the world which they both share as their objective is the "stratigraphic analysis of reality,"[6] Calvino believes that science should also be taken into consideration since it also deals with problems of the nature of the world and offers ever new interpretations of it. He declares that, "We will not have a culture equal to the challenge until we compare against one another the basic problematics of science, philosophy, and literature, in order to call them all into question."[7] This he sees as occurring only in a distant future.

And although he considers the deeply emotive nature of culture, which he calls "erotisation," to be playing between signs on one side and meanings on the other, as well as between myths and ideas (the one side being evocative, suggestive, which side galvanizes our attention and calls upon all our natural endowment, and the other side being intellectual, which side opens horizons of imaginative thinking, speculations, projections, and perspectives of thought) and that consequently a philosophical aura is present in most of the great literature (Aristophanes, Voltaire, Defoe, Beckett, Gombrowicz, etc.), it all ultimately, for him, boils down "a relationship between the phantom lightness of ideas and the heavy weight of the world."[8] Equipped with these two different ways in which to carry out their investigation of reality, each opens its tent.

So, Calvino holds that literature's investigation of reality is just one of three possible ways. The third partner which Calvino brings into the enterprise, namely, science, has an indispensable role to play. Here Calvino falls into a trap of his own making; or, to put it another way, the futility of his proposal comes to light. What could the role of literature then be other than to present us a positive truth about the state of affairs in the world, the same kind of "objective," universal truth that philosophy is called upon to discover and that science aims at? This would be a much more thorough-going type of intellectualism than that of the reflective writers mentioned above.

We emphatically reject such a view. The role or vocation of literature, whatever its affinities with philosophical concerns and issues is unique. Its role is not to present us with the cognition or knowledge of anything (e.g., the world, man, life, etc.) but to take up issues that lie at the much deeper level of human existence. The role of literature, that to which its means are geared, is not to explain the world and life as we discover it by positive, universally valid, intellectual means. It is to *re-create* the world and life after we have already lived it and come to know it in the positive sense, to transform what trivial and bare positivism yields through the *creative vision*. Furthermore, although it is correct to say with Calvino that the material with which both philosophy and literature deal is language, that is, the word, meaning, syntax, etc., and that it is different use of this material that differentiates them, this is only a most superficial statement. Those writers who capitalize on a mere mastery of language and make style their main objective never reach even the surface of the foundation of great literary works. The writer does not merely seek to express himself — to equate the accomplishment of his task — in a new form or technique, in the manipulation of tones, periods, vocabulary, etc. Literature is not an act of virtuosity. And the use of language that the literary author chooses is not meant to convey mere information, or knowledge, as is the case with philosophy and science: it is meant *to establish contact, to convey experience*.

But neither, as Ernesto Sabato tells us, does the artist "traffic with his emotions" in order to concoct from them representational figures, situations, etc. What the artist seeks, and we can accept it on Sabato's testimony, is to say his "truth," that is, "his vision of the world."[9] It is because they embody this vision that the heroes of a novel impose themselves upon the imagination of the writer and are saved from the flood of the turmoil of the life-stream. They become his heroes insofar as they emerge within his sustaining vision of human destinies. The vision essentially "defetishizes" the temporal dimension of everyday concerns, struggles, striving, victories and defeats. This vision does not describe or investigate, but recreates the world, recreates life.

These are crucial points on which philosophy and literature diverge in their roles and in their respective vocations.

(b) *Conveying a Vision*

To sharpen our argument, let us return to the creative vision. When we go to the creative crucibles in which both philosophy and literature originate, we find that although both are creative endeavors of human beings each of whom possesses a *creative vision* (to which I have devoted ample attention in my previous writings),[10] they take different routes. Philosophy aims at discovering the positive truth about life, human existence and destiny, and the universe within the compass of the lines that the creative vision draws. Literature, however, does not seek knowledge of the positive state of affairs of life; on the contrary, it dismantles that state of affairs and re-creates it in a unique fashion in order to convey the unique *personal* vision of life which the writer summons up. Philosophy seeks to establish its findings in universal, objectively valid, rational language. Literature — the novel, in particular — seeks to incarnate a vision of life, human existence, etc. in a *deeply affective* crystallization in a body of writing which after having by appropriate means captured it, will be able to convey it to the reader "in person." That is to say, while philosophy aims at establishing rationally a schematic pattern of descriptions, structures, and their laws in order to provide principles explaining the definitive nature of reality according to a most general outline of the vision of each philosopher, the writer focuses on his vision itself and coaxes it from the hazy, undetermined, limbo-like state in which it first floated before his "inner eye" in order to give it the most particular, specific, personal incorporation in human life-situations, characters, intertwinings of events, etc. in accordance with his deepest feelings, emotions, strivings, and urges — stemming from his own flesh and blood, and spirit.

Yes, both types of creativity work with words and meanings. However, the philosopher chooses for his purposes, if not always the most abstract and universal words, then certainly those words with the clearest denotative and evocative effect, so that a common consensus about his meaning may be best assured; he besides presents his vision in a schema of impersonal, clearly connecting — or disconnecting — unifying links to achieve universal validity. An epic work of literature, on the other hand, states its case in the most striking fashion, one that touches the sensibility of the recipient at all the appropriate levels of feeling, emotion, response. To this effect, the author chooses some or other "plot" of events to get the recipient involved in a life-simile, and it is by being carried along by its course that this most concretely crystallized vision of the writer can be lived anew by the reader as his/her sentient endowment spills over into the realm of the spirit within which the philosophical overtones of the work will appear.

It is proposed here that it is philosophy's role and meaning to give man, through the exfoliating of its vision in a systematic, rational appeal to universal and detached thinking, an understanding of reality and to enlighten him, intellectually and objectively, about the parameters of human existence, its nature and prospects, options and limitations; it offers this clarification to all men in all situations and also indicates the proper conduct for their fulfillment. Briefly, its significance is to bring to man enlightenment.

Literature operates in a reverse manner; the most personal vision of the writer does not appeal to the most lucid zone of the reader's thoughts but seeks out the remote regions of his/her thought, the emotional climes where in obscurity existence is carried on its way, even though the thoughts held there are tamed, and often so dominated, and their voice so muffled, that they may not reach the surface of everyday life and sensibilities — which are thick with the deposits of personal, societal, and cultural influences overlaying our deepest self.

The author's vision eliciting a specific feeling about the personal life of mankind aims to be partaken of as "life"; the language which the writer chooses is not meant to propound objective, universal truth, but *to establish contact* between the living reader and his vision of life. Hence, the drastic difference in the use of the "raw material" of language as Calvino calls it. Hence, the dilemma which the existential novel finds itself facing, for while it attempts to bring in philosophical ideas too clearly, it is not in position to make its language vibrantly compelling.

Yet what would be the upshot of the establishment of this contact? These reflections bring us to the philosophical core of the present investigation: *the subliminal logic of the significance of life* out of which the literary work springs and in which it then fulfills its vocation within the reader.

We will endeavor in what follows to advance in our investigation, — already initiated in our earlier work — of that subliminal forge in the human soul within which the life-significance of the work of art — here, literature — is ciphered by the writer, as well as de-ciphered by the reader/recipient.

CHAPTER TWO

THE VISION OF THE WORK OF ART AND THE TWILIGHT ZONE OF CONSCIOUSNESS

INTRODUCTION: THE ORIGINATION OF CULTURE AT THE POINT AT WHICH THE VITAL AND THE CREATIVE FORCES ARE LINKED

(a) *The Roots of Culture*

Aesthetic, artistic, literary creativity has significance for human life insofar as it not only enhances our otherwise strictly pragmatic, pedestrian, and trivial approach to life by responding to our deepest longings, nostalgia, and strivings beyond concern for survival but also stimulates our creative imagination — and thereby our entire life-oriented functioning — to undertake new approaches to basic human needs even as it responds to longings and dreams, so that life/survival concerns intermingle and fuse with the deepest human concerns, that is, with the concern for the aesthetic significance of life which only humans seem to manifest. It is this intermingling, fusing of both their reciprocal prompting and exchange of influences, that constitutes human culture. Indeed, the role of bringing into the common pool of experience the specifically human factor of *elevation* above strictly practical concerns, of expanding the range of our experience through *sublime* sentiments, of shedding the glow of *beauty* upon an otherwise aesthetically neutral view of the world and of life, is one that the arts and literature alone perform. The infinite variation of aspects, shadings, and types of aesthetic sentiments that accompany these three moments constantly suffuse our valuation of life. It is impossible to envisage our world of life without them; without them, that world would lose its human significance; it would have never unfolded into a cultural world.

It is precisely culture that is the key to our knowledge of human history and past human societies.

If art and literature fulfills this role with respect to human existence it is due to the fact that in their origin — that is, in the origin of their enactment and in the awakening of their reception in the spectator or reader as well — they penetrate to the *Archimedean point* from which

A-T. Tymieniecka (ed.), Analecta Husserliana, Vol. XXVIII, 21—31.

human life takes off, that point at which our animal functions and forces receive the impetus as well as the resources which the onset of the Human Condition brings. This point is where the creative orchestration of human functioning gets off the ground, the zone that encompasses the functioning of what I have called elsewhere, the "creative forge," a phrase familiar to the readers of my work. What I plan to explore in the following pages is the nature of this zone and its uniquely strategic position with respect to the unfurling of the progress of human beingness. In fact, we will see that the reason why the creative endeavor of the human being as exemplified by works of art and literature, brings the uniquely human factor of life-significance into the human undertaking *in toto* is that it proceeds in its operations *from this very zone where all the resources and tendencies of life's unfurling meet — there it is that life's significance is to be ciphered.* Conversely, in order that the artist may establish *contact* with the recipient so that he/she may *receive* his/her communication, the recipient has to dive to the same Archimedean point of his/her beingness in order to decipher that significance.

Our purpose in exploring the zone of the "twilight of consciousness" is to see how this exchange, which decides the unique vocation of the arts and literature, occurs.

(b) *The Subliminal Forge of the Soul*

We have elsewhere proposed the conception of the human soul as a territory of life that encompasses the empirical psyche, not only the dynamic circuits of the conscious life but the pre-conscious as well. We have emphasized in particular the zone of the soul in which the transition between the two takes place; for without that transition the empirical would in its encounter with the mind's intentional/conscious circuits lose its significance without acquiring a mode indispensable for entering into the cooperation with the mind; for the field of this encounter is the subliminal circuits of the soul's functioning.

This zone of empirical circuits of functioning is neither conscious nor pre-conscious/subconscious, nor pre-predicative, for all of these have already a reference to the mind, to consciousness; it does not partake in itself of the mind's light; the primal operations of Nature remain in "obscurity," in "darkness." The rays of intelligence which penetrate it come from the workings of the creative function of human beingness; because this is the zone of the play of "light" and "darkness," of the elementary forces of Nature and the creative intuition of the human quest, from which play proceeds the specific line of cultural beingness, we will call it the "twilight zone." It is this twilight zone which is the subliminal forge of the human significance of life that the work of art originates as the creative vision emerges into the full light of the life-world and illuminates its objects and to which it returns in the recipient's soul.

Now, turning directly to the investigation of the place which the twilight zone of consciousness occupies within the conundrum of the human soul, let us observe that often, as in this case, what comes first in the line of generation is last in the line of discovery or apprehension, and vice versa. And so we have to begin our descent into the primogenital workings of living human beingness at the level which comes last in the line of its unfolding but which is primary in the manifestation of the human conscious/pre-conscious/primal apparatus. So, of the tangle of circuits of the life of the soul, the one which presents to us the world and ourselves, as if already made, is the organized circuit that emerges through the work of the conscious apparatus which is penetrated by the intelligible sense and directed by the constructivism of the intellect, the agency with overall extensive adherence to the central translucent self. It is this circuit which manifests itself as our organized field of intentional consciousness in the investigation of its poetico-noematic structures undertaken by Husserlian phenomenology. However, this circuit which establishes the conscious-luminous zone of psychic life emerges as a projection from and above a pre-noematic, that is, pre-intentional and pre-intellect-structured scheme. This latter circuit, subjective — as it is called — is already "pre-objectivized," that is, with respect to it the intimate inner realm of the self, has already assumed its universally valid posture, "direction toward" objectivized forms, and should in fact be considered to be already the "intersubjective" circuit of interiorized life.

This circuit, upon which — and from which — is singled out material to be used in the constitution of objective forms, plays the role of an intermediary zone; this zone lies just below the universalizing dominion of the lucid intellectual intentionality that prepares the setting up of a subjective/intersubjective life-world in which conscious communication between individuals is the crucial factor for the establishment and progress of a uniquely human *community of life interests and accom-*

plishments, and just above the primal zones of life. It is a zone which hangs in mid-air as a chaotic depository of material, of vaguely designed shapes, truncated fragments of memory, images of thought, flashes of faded feelings and desires, etc. from the longer or shorter voyage in time. It is upon and from within this chaotic, more or less passive/lively, turmoil that the powerful constructive work of the intentional system moves. This material gathered from memory, coming back at random, and from the stimuli of the present moment from which fresh experience surges, is already the half-prepared ground for the universalizing integrative work. This latter establishes our simultaneously personal and public world. In this zone, the internally originated personal associations of truncated fragments obtain a proper place within a new intersubjective association; if they are not needed, they recede back into this intermediary zone. It is this zone that psychologists have considered to be that of the "subconscious," which itself offers the reservoir of the unforeseeable resources of the psyche. This psychic background of our lucid functioning may often — beyond the control that the intentional agency exercises — break through the constructively organized intentional network with its own tendencies, its own dominant themes, and project or move on its own into intersubjectively oriented active reflexes.

Yet this zone of circuits is, as mentioned above, merely intermediary and itself hangs above the truly *primogenital zone of the primal life in which the soul has its roots*; its circuits draw forces from the womb of Nature; its rules and processes promote the organic and vital order of life. This deepest, darkest zone — because unpenetrated by the light of the intellect — this "abysmal" zone — because once we venture a "plunge" into it, it seems to draw us down to a bottomless depth — seems to offer a terrible challenge to the intentional power of humankind when we attempt to "bring light to it," to "clarify it" directly. However, artists, writers, philosophers insofar as they are creators have roundabout access to it through the creative function. Following them we may investigate it in its unfolding of the type of beingness that is human.

While all our constructive specifically human functions fall naturally under the aegis of Imaginatio Creatrix and aim at the differentiation of living human beingness from the universal current of life, this constructive work has to be accomplished within and in cooperation with the organic and vital circuits of our brute forces; these forces which belong to the wild onrush of the universal stream of life are bound up in the circuits of the dark zone of the soul. There is where they have to be met and tamed. Already part and parcel of the individualization of life, they have still to be tamed for the sake of the specifically human avenues of life and the unique modalities that Imaginatio Creatrix inaugurates. So there appears to be three interrelated lines — the aesthetic, the moral, and the intellectual — and each of them plays a distinctive role in the working out of life-significance with nature's organic and vital forces serving to sustain the existential development of living being. These forces are resistant to entering into novel configurations, to entering into constructivism at all; they must leave accustomed paths to go beyond simple organic-vital functioning and the most basic individualization of living beingness; these primogenital forces react especially to all efforts to expand their range to constructive avenues, lines, sides other than those promoting the vital sense of life. And yet when the human condition enters into play, all constructive functions unfolding human beingness fall under the sway of the creative orchestration. As I have pointed out often, the creative orchestration of human functioning takes direction and control of the entire range of our functional circuits. Yet in order to take it, it has to tame the organic/vital circuits and bend them and mold them in its particular manner, drawing them into its own stream. It aims at the establishment of means for the stabilization of the inner/outer flux of life's force, and seeks to distill out of the relentless rush onwards in all directions some factors from which can be built a raft of relative repose to throw upon the stormy waves of life's flood; a raft, the fabric of which separates us from the engulfing urgencies of the flood that will not of itself halt to give us a chance to differentiate from it our own course and would otherwise work itself through us; this fabric gives humankind a relative hold upon the flux such that we may manage a network of our own controlling devices, while the primal zone of the soul, following the stream of life, challenges and repels the effort to be grasped or elucidated in terms of any stabilizing devices. Yet this zone challenges these attempts by the lawless turmoil of the ever turbulent motion of its innermost pulsations which transmit those of our physiological process, its blind drives having no aim or target.

When confronted with an intuitive plunge into its workings, the primal zone resists and reacting explodes in fear, anguish, pain which pierces the screen of our intentional/conscious domain, invading it.

This zone is evoked mainly in psychopathology and is unduly confused and identified with the pre-conscious or subconscious. In fact, in pathological circumstances, a destructurized consciousness — falling to the level of the pre(sub)-conscious, is abandoned to and preyed upon by this primal circuit of the life of the soul. Yet it is into this same realm that the creative genius plunges to harvest its greatest resources; the opposing notions of devil/angel, witch/saint, hero/scoundrel, dragon/dove, etc. are, as we will show further on, the fruit of the creative genius of humankind's dives into this abysmal zone of the soul to install there its *subliminal* workshop. *In this creative circuit of transformations, Nature becomes humanized.*

Let us, however, begin with the disintegrating descent to the primal soul. In the psychotic destructuring of the conscious faculties, the prevailing natural, defensive tendency to not dwell upon the truncated fragments and the deformations of objective forms which enter into haphazard association no longer prevails; the stereotypic links with which the universal ordering captures the products of the abysmal realm are then ignored so that the veil which the structurizing of intentionality throws over it, as if "exorcising" the frightening tentacles which threaten life's poise, is lifted, and then through the cracks of the dissolving structure comes the pull of the stygian vortices that drag down into chaos, oblivion, unreason, anguish, and fear. No longer enlisted by a stabilizing tendency into coalescing in works of solicitous cooperation and construction, these formerly fragmentary terrors erupt in a blind violent outburst against everything, as if defiance and assault on all about were the only way in which this territory of vital subsistence could defend its interests.

As we see, the constructive intentional progress is indispensable for the maintenance of the universal progress of life, and without occupying a position/place within the time/space life-world, an individual loses his vital balance and is incapable of holding on in the life-stream on his own.

Do we talk about this "abysmal" groundwork of our psyche only in a hypothetical way proper to science, or do we not individually get at times an awareness of a glimpse of an insight into it? Although it came into fashion with Freud to take the "subconscious" into consideration, there must have been as long as humanity has existed awareness of the profound stirrings of our psychic fountain which we may, in fact, if not directly "see" or grasp in an objectifiable way, since it lies below the threshold of the lucid zones on the surface of our field of consciousness, at least experience vicariously as being below that surface. Certainly, in various circumstances — in situations which I have termed as being "on the brink of existence," when the texture of our safe, stable, reliable, and peaceful everydayness disintegrates under pressures which we are not able to counter — we have always been confronted with the "underground" turmoil of our inner self. When we feel, in fact, our edifice of world-relating links and concatenations disintegrate and our seemingly hitherto well-established life-connections with others loosen and fall apart, we are not in a position to remain at the "surface" in the lucid conscious sphere in which the spectacle of the world has always appeared with ourselves included in it. As a matter of fact, this spectacle itself, under the pull of the "deeper" forces of our vital system, which have slipped the reins of constructively oriented "taming" structures and are now "wild," recedes from our full experience, its uncoordinated fragments pulling further and further from each other until it vanishes altogether. Our attention which was hitherto bound by the constructive projects of our constitutive lucid awareness loses its hold on the realities of fact and becomes disoriented, as it turns to glimpses of our inner self. We seem to find before our attention bits and pieces of images, ideas, thoughts pointing out to some or other intricate development; we are drawn into their labyrinth down innumerable paths and through doors that open to them; we are enraptured by the subtlety of the hazy issues and the delicate delineations in their interconnections which invariably are lost from sight but which open new fascinating perspectives. We have to do violence to our attention to bring it back to reality which appears then to brutally cut through, the riddles of delicate issues now appearing to be simplistically and crudely presented. The actual world is then dwarfed, and the knots of feeling which run through the heart appear rough edged being drastically severed. It is in the sensitive universe of an exceptional person endowed with a feeling system comparable to the strings of a lute or of a harp that we enter into the subliminal circuit of our beingness, and the person who then appears is hardly comparable to the life-enlightened self whose skin is thickened by struggles and needs to not hear, not see, not be "overaffected" by pain, humiliation, sadness, etc. as it goes about the business of life. But in those rarest of moments when we delve into our subliminal spheres, the real-life person who operates in the translucent world of life appears to us to be only a phantom cut off from the springs of the beautiful, the gentle, the caring, the sublime. We join this deep self of ourselves only in chosen instants of reflection, of particular serenity, and also in those instants when before falling asleep our psyche seems to plunge into itself and retrieve from its quiet waters some glimpses floating there of another life sunk in its depths, an infinitely richer and enhanced life ready to envelop us. We naturally fix our "inner eye" upon one or another image as pointing through some unclear associative links to bits and pieces of memory which are somewhat lost out on a vast and foggy horizon. However, even here and now, in the life of actual facts, we may witness a scene and "recall" it, as if it were a double of some place or time already visited in a dream or premonition. Our attempt to recall the occurrences of the dream draws us deeper into ourselves opening a labyrinth of hazy perspectives with strange inter-linkings, each of them drawing us further into some mysterious dealings which our psyche appears to conduct as if "behind our back," beyond our control, "somewhere" at the very depth of things.

It is into this third, abysmal-vital, zone of the soul with its primal pulls and eruptions of forces that imaginatio creatrix, while establishing the creative orchestration of our constructing faculties for the sake of life's animal and human progress, sinks its roots.

(c) *Fabulation as the Primogenital Factor of the Creative Orchestration in the Initiation of the Human Significance of Life: The Paradigm of Life and Its Dramatization*

When the creative orchestration under the promptings of Imaginatio Creatrix sets out to establish the subliminal forge of the human significance of life within this "wild," vital zone of the soul, it simultaneously organizes itself into a system dealing with otherwise opaque and numb rebellious forces, drives, etc. In order for these to be "tamed," that is, brought into the common constructive pool, they cannot be met and treated in their infinite singularity and variety. Within the range of vital circuits, as they surge and manifest themselves within our human organic existence, these forces emerge in undifferentiated complexes and "mean" just blind reactions of pain, pleasure, hunger, satiety, etc. Their "meaningfulness" is rhapsodic; it partly signals the malfunctioning of organs, partly manifests the state of affairs in the individualizing development of animal beingness with reference to the appropriate criteria of its species and the universal system of life, and, lastly registers the challenges which situations present. Even in the higher animals endowed with a psyche (e.g., dogs, cats, horses, apes, etc.), the meaningfulness of their reactions and of their appreciations of them seems to be restricted to the singular life situation in which these reactions occur; this appreciation is also narrowed to the immediate survival concerns of the individual and expanded merely by reference to the more general pattern of survival/comportment.

When the creative orchestration settles in, this meaningfulness is truly expanded; meaning is not simply found through the dissociation of the complexes that enter into the primitive instinctive vital reactions, although dissociation does come into play. For in order to dissociate these complexes a set of a prior available principles is indispensable, principles coming from general "views" already "floating in the air" which serve as points of reference for the differentiation of the moments of a complex in accord with some "reasons" for which the coalescence occurred.

Without entering here into further analysis, let us just note, then, that before — or together with and yet principally operative within — this encounter with the primal forces of Nature within us, the creative orchestration is provided with provisory schemes or "views" which allow the treatment of these forces; these schemas are the result of the response of the creative imagination in this encounter. It is imagination that within the entire creative orchestration as it draws on the three primogenital sense-giving factors — the aesthetic, the moral, and the intelligible sense — and that, responding to the urgencies of life transmitted by the primal forces, projects a range of appreciations in its *own* fashion, *a specifically human fashion.*

Here we are hitting at the crucial point in this entire enterprise of "humanizing" the vital stream of life. In fact, the creative orchestration has a primordial function around which all the other sense-giving functions revolve: that of fabulation.

Indeed, with the advent of the creative orchestration, our living beingness follows a path with its very own initiative, objectives, and expectations; the human being projects into the vital interest of life *his own interests.* That is, he has to have general appreciations, first, of his own situation, second, of his aspirations and prospects. *These views are provided by the imagination as it fabulates.* Instead of seeing each physical, practical incident of life as a one-sided stimulus-response event, imagination focuses already on the effect of nature's (physical) gregarious promptings, envisaging them with reference to the entire self-individualizing project of living beingness. Then it grasps these encounters between the primitive forces of Nature and society within the individual in terms of the "human predicament." Furthermore, fabulation universalizes their significance by seeking paradigmatic features of these recurrent encounters. Lastly, it models them into sequential hits heightening the role of some moments, one over against another, sharpening some contours over against others, etc. In other terms, there occurs in fabulation a basic and *uniquely important dramatization* of the events and their effects, a modeling of views about human predicament.

Fabulation, whether expressing the sequentially grasped (formulated) human predicament by enacting it in dance, ritual, mimicry, a stage play or by way of discourse in storytelling or writing, revolves around three basic principles: the tragic, the comic, and the epic. Each of them performs a unique role with respect to the life-significant task which it is called on to play in individual and societal life. Myths, rituals, sagas which are formally crystallized in a dramatized paradigmatic view have always been, in whatever form taken from culture to culture, the first guides in a culture's history.

And it is precisely with reference to the fragmentary significances drawn from the paradigmatic grasp of the human predicament that the primal forces of the soul are molded in their encounter with the creative orchestration. The struggle for human life, for survival is now consciously assumed by our living beingness and oriented by insights and strategies which it devises and enacts. The paradigmatic fabulations project antennae for us by which we form appreciations and devise our strategies.

In the encounter with the primal forces the cooperative responses seek to coordinate them in the constructive project of living beingness, and so they acquire a molding tonality which is propitious for the emergence of the specifically human sphere of life: comfort, repose, hope, enthusiasm, benevolence, attractiveness, beauty. The rejection of the constructive tendencies in human experience imparts an imaginative tonality of all that is life-hindering: anguish, anxiety, dread, greed, malice, malevolence. Furthermore, the dramatic heightening (stringency) by moving between imaginary extremes introduces an emotionally charged dialectic of "infernal" forms of life's destruction and corresponding "life-beneficent" forms, poising them in a life/death struggle; a struggle which means in the extreme the confrontation of the striving elementary forces, antagonists or partners in the quest for a new humanly significant equipoise which is indispensable for the equipoise of life as such. The malefic "infernal" forces conjured from the blind recesses of Nature appear on the human platform and make their stand; they take on forms according to some human measure and may thus be exorcised of their malefic strain by gifts, sacrifices, exhortations, and lastly by confronting the exaltation of their opposites; and then, as if tamed, they are brought into balance with the positive tendencies and constructive demands of life.

Evil, and its opposite Good, seem to govern not only the moral life, but to be projected upon the ultimate forces of the universe: the divine good and satanic evil, which emerge from opposition, stem from the same primal zone of the human soul as the source of the ultimate significant principles, principles which setting up points and splitting off — or extending to the antipodes, of human fabulating life-predicaments, one pole of which is well grounded in its concrete base within the system of experiential objectivity and the other of which in corresponding vitally rooted pre-experiential irrealities — sustain in their tension the relatively stable platform of humanly projected existence. Myth, legend, religious mysteries, dreams all find their origin within this aboriginal contest between the relentless impetus of life and the opposing individualizing constructive tendencies which struggle to capture it in coherent structured operations with directional vectors and a rationale that can process its forms into a stabilizing machinery and channel its flux. Myths, legends, mysteries all express this limit situation — the limit zone of human existence with its innermost struggle for survival-equipoise, the struggle to maintain itself on the surface of the flood while in some indispensable measure controlling and dissecting it.

CHAPTER THREE

THE RECREATION OF THE VISION OF LIFE BY THE RECIPIENT OF THE WORK OF ART

(a) *Everyday Life-experience*

Let us emphasize again that literary work is certainly not the instrument of choice for transmitting to the reader a rational view patterned and arranged in orderly schemata reaching down to the first principles of an explanation of why the human world moves along its course, why the human being is caught in his situation within Nature and society, etc. Clearly it is not fitted for giving such an explanation. While involving the reader's innermost feelings in the proceedings of life which it presents, the literary work aims at his/her experiential horizon to explode beyond what everyday practice with its narrow practical frontiers allows for. It aims in the first place to unfold within the reader secretive, inert, waiting bundles of emotions, concerns, forebodings, anxieties and bring time to the surface; there to ignite them with the magic touches of imagination brought to it by the writer, and to elicit from their resources further emotive pulsations within the psyche of the reader illuminated by the inward "light" of the work's creative vision, which presides over the literary work throughout.

In the second place, to bring these freshly illuminated emotions into the experiential molds, which by their wealth of novelty break the old ones of their power is to call for new molds. Lastly, experience in its expanded molds reaches the current life.

When we propose that the life significance of literature is in this way a special transmutation of the reader's experience, we must then ask what this everyday experience consists of that it be prone to such treatment; and, why does it need the intermediary of the creative work, the guidance of the writer, to resonate? To this effect we must briefly sketch the nature of everyday experience at large.

In beginning with everyday life-experience in dealing with experience, we already deal with the human being who as such is in the lucid zone of his functioning and is a creative being; he has risen already from the brute functioning of his animal vital circuit by implementing creative imagination as the principle orchestrating his functional schema

A-T. Tymieniecka (ed.), Analecta Husserliana, Vol. XXVIII, 33—44.

and, simultaneously, has through the operation of the newly surging faculties of memory, will, and the intellect established the self-interpreting agency of his mind in the basic act of experience. That has already transformed the brute and numb and blind moves of vital nature into complex operations through which the objective world around us and ourselves as agents within a community of others like ourselves emerge. That is, our experience already binds our natural stirrings and moves into innumerable networks, molding them thereby, establishing lasting molds for the repetitiveness of experience, this latter being the indispensable foundation for the efficiency of our ever objectifying life-progress. This constitutes the already "humanized" zone of Nature within us. And yet it does not bind all of the stirrings of our vital nature. There remains beneath the threshold of the activated experiential zone the elementary level of our soul which belongs to Nature within us; pure Nature through which all the organic and vital forces of flesh transpire, pulsate giving us Nature's force and vigor and spontaneous urges to move, to advance, to rush forward.

But although we are already "humanized" and have, furthermore, developed in our experiential progress cultural molds for experience which constitute the specifically human zone of the soul — the subliminal or elemental sphere of the soul — nevertheless, our everyday experience does not keep this subliminal soul in an active, expanding, flourishing state. How could we focus constantly upon the specifically aesthetic, moral, intellectual aspects of our feelings, emotions, thought, and reflections — and upon those aspects of our conduct, as well — developing, refining,and intensifying them according to our taste, which itself would then unfold in practice, when to live means to perform, of crude, vital necessity, a series of repetitious everyday tasks *ad nauseum* as well as to engage in numerous life-transactions with others, transactions upon which our survival depends and which totally occupy our acumen. This means that our everyday life-course imposes clear-cut simplistic patterns upon the exercise of our life and confines our experience to rudimentary, simple forms. The subliminal expansion of the soul is in everyday life constantly dominated by vitally significant, societally required urgencies, and consequently what comes to the surface in everyday experience is robbed of all its subliminal aura. Our practical tasks and rounds and the urgency with which they must be accomplished very rarely give us the freedom to bring forth in their accomplishment our deepest subliminal tendencies, aspirations, commitments, and reflection. This reflection is in the current of life dismissed as "useless" or even a hindrance to its progress. Rushing relentlessly onwards, we remain out of breath; our deepest subliminal aspirations and convictions remain hidden behind the currents of everyday concerns. They lie there in the twilight, obscure to us, who then act half-blindly, not knowing why. And yet, it is in this twilight zone that the key to our deepest self lies, the key to our "understanding" who we are and to the exfoliation of our destiny.

Do we have to be ever carried along by the aimless current of life, helpless and lost, or do we seek some enlightenment for ourselves? Presuming that we do, I submit that it is art's vocation to give us a helping hand and be for us a beacon.

(b) *The Fountains of Culture: Ideals, Virtues, Sacralia, Taboos*

The question by which we terminated the preceding section launches us into a hyperbolic line of argument. In fact to show the initial stage at which the recipient-reader-spectator retrieves the gist of the work of art — of the literary work in particular — that is, partakes of the life of significance that the entire sequence of a specific cultural branch of human history has deposited in the work's original form as given to it by the author within the intentional form of the spirit, it is necessary to evoke, be it ever so briefly, the evolutive trajectory of a culture from the incipient take off the human spirit from the bowels of Nature to its high point in a flourishing civilization. Reviewing these stages in every great civilization, we find that it is law-giving, art, and literature which punctuate their progress; they are its midwife as well as wet nurse — and its manifestation too.

There is nothing new in this statement; it seems almost to be a banality. An essential contribution can still be made to the discussion, however. By raising the issue within a new field namely, that of phenomenology of life, we are in a position to give its ultimate clarification, and thus, justify this otherwise vague claim. We can do this by tracing the creative outburst to the level of the Human Condition seeing it as its very crystallization. Art and literature appear as crystallizations of the answers to the deepest longings of the subliminal soul, thereby, responding to the needs of the human spirit as it takes off; they compensate for and complete human limitations.

The human being falls short on all sides in his forces, capacities, means, wits, powers, judgment, and talents in the urge to respond to his innermost call; he fails to become his own master within the intricate web of life, and longs for completion, compensations, reward, and redemption. To meet these longings, art in all its forms gives us ideals, dreams, beauty, virtue, sacralia, taboos, glorifications and damnations, etc. Bound to the earth by his feet, the human being requisitions the open sky in which to fly upon the wings of the spirit.

To substantiate this statement let us refer to the evolutionary curve of the culture's unfurling within our own perspective, that of the phenomenology of life. That is, let us return now to the crucial role of fabulation.

(c) *Wonderment and Fabulation*

The ancients considered philosophy to be born from wonderment. Wonderment appears, indeed, to be an essential stance of the human being, and this not only when philosophical reflection emerges in the unfolding of the human spirit. As a matter of fact, wonderment in its primitive context, wonderment about things which do not seem to fall into the simple causal concatenation, has its experiential counterpart in marveling; one completes the other. When we become "enchanted," entranced before the beauty of a flower, plant, tree, this is marveling. When this beauty or gracefulness, or complexity makes us wonder about the prodigality of Nature in bringing about forms, and about this very bringing about, this is wonderment.

We could say that the marveling-wonderment coupling is the primogenital condition of the launching of the human spirit. That the investigation of the primitive societies shows that the most primitive societal organization follows intuitions of the sacred illustrates this point. The intuition of the sacred's entering human empirical, survival-oriented existence is the outcome of the marveling-wonderment experience. With that intuition, the human being queries beyond strictly practical interests. Our basic physical nature fails to supply understanding in this quest. And so, to complete this quest, the human spirit initiates culture.

However, the primogenital human stance of the marveling-wonderment experience is not rooted in a simple streak of functioning; to the contrary, it is an amalgamation bearing a triad of intellectual, propulsive bents of the human mind: *natural curiosity, pure inquisitiveness, and personal conjecturing.* Each plays a part in this experience. And the surging of each is complemented by the mind's receptivity to the "unknown," the "incomprehensible," and by its reaction to all as a "challenge." It is the activation of this latter propensity that is the necessary precondition for the marveling-wonderment.

The act of attention, which in agreement with Husserl, Minkowski, and others, we have previously presented (See *Logos and Life*, Book 1) as the very first movement of the intellect, is still upon this ampler analysis the primordial act of the intellect. It is responsible for the initiation of the constitutive — intentional — genesis of the structuration of reality. Nevertheless, *it is made possible and activated by this triad of propulsive propensities and the primary reaction of the mind to stimuli in terms of a "challenge." Natural curiosity, pure inquisitiveness, and personal conjecturing are the primary circuit of the intellect's movement.* With this circuit's activation, the fabulating function of the human mind is awakened and called on to process imaginatively and creatively the material set before it, to select from, enhance, embellish, magnify, embroider upon, etc. the human condition — on a canvas of sky as it were.

Joseph Conrad, shows rightly in his incomparable inner/social drama, *Lord Jim*, that "facts have no meaning." Without imagination and reflection, they simply remain meaningless. We otherwise do not "think" about them; when confronted with them we simply take instinctively the most expedient course of action that we are capable of. We endow the brute facts of life with significance only as our mind moves from marveling to wonderment to fabulation. With this coordinated progress, the "individual's acculturation" from scratch began in the pre-cultural epoch that now lies in oblivion; and so acculturation advanced from generation to generation in complexity and refinement of the spirit. The progressive recourse to and the full emergence of culture comes about as fabulation chooses its various routes and modalities: ritual, dance, mime, play acting, song, stories, the plastic arts, the organization of societal life, laws, principles of distributive justice, etc. In fact, to sharpen the issue, the deployment of fabulation, in which events, conduct, the particular human predicaments of the period, the attitude of the human being toward the other, as well as to the group, and the many challenges which the human being encounters and has to face in the struggle for survival are all scrutinized in their brute reality and then imaginatively transformed into significant phenomena, has promoted and continues to promote the modalities that people's attitudes toward one another take within the group as well as the modalities of the attitude of each individual toward his life course. The few documents that survive from Occidental culture's remote times show how events occurring within human communities were not passed over in "silence" but were seized upon by bards, priests, soothsayers, minstrels, prophets, and glossed on in marveling and wonderment.

Wonderment at the unforseeable, the unavoidable, the incomprehensible, at the enigmatic maze of otherwise factual occurrences stirred and fed the fabulating function. All such causes for wonder offered it topics upon which to dwell. Human fears, needs, dreams release the latent propensities of the subliminal soul, and to respond to them the fabulating imagination sets to work. Opaque events, situations, the people involved in them are imaginatively "de-composed," taken from their dense factual interrelations, and *conjecturally* transposed into a transparent pattern of relations and embroidered upon. It is in the interval between this decomposition and conjectural transposition that the factual occurrences, having been plunged into the creative forge of the Human Condition, direct the creative imagination to reach into the conundrum of the life conditions which confront the human being as its existential predicaments, predicaments which are inherent to his struggle for survival. At the same time, the work of the creative imagination stirs the subliminal soul to imaginatively react to these vicissitudes. That is, as we have expounded elsewhere (Cf. "The Aesthetics of Nature in the Human Condition," *Analecta Husserliana*, Vol. XIX), the brute and opaque facts of life are imaginatively transfigured into *elements of existence*, and to confront them there emerge *the elemental passions of the subliminal soul.* These passions serve as the bastions of life's struggle and the beacons for its orientation. In the second part of this treatise we will analyze light and the sea as two elements which evoke/reflect a host of corresponding passions. Here it is appropriate to emphasize the continuous work of creative imagination in making the crucial passage between nature without and nature within man. This passage takes place precisely in the phase of transposition. In fact, with this transposition the facts of real life lose their matter of fact concreteness; they are extracted from the opaque conundrum of their situational entanglements. They appear now in an imaginary network of intricacies which are, however, as conceived by the fabulating imagination, apparent. This apparent network makes the situation of the facts, people, and events appear ampler; that is, the intricacies that make up this picture enlarge the available possibilities in life. The essential patterns of the entanglements of situation and event are now related to a pattern of the personalities of the people who are caught in them but at the same time can or cannot command them. In this pattern of personalities, specific bents of character, emotional inclination, etc. are revealed.

Within this new creatively invented pattern, situations and personality acquire universal validity. Patterns so projected are further expanded through the links between the possibilities which they contain and the recipient's emotive propensities, expectations of the future, and relevant hopes.

The foremost source, however, of the new links introduced after the intentional dissociation of factual reality from the opaque conundrum of the life setting are the three primogenital sense-giving factors — virtualities — of the Human Condition, which find in fabulation's getting to the ground of the spirit their main occasion to unfold. The *poetic sense* is at work in the expansion of the experience of the spectrum from the beautiful to the ugly; the *moral sense* brings to human relations the measure of the benevolent and the fair and their opposites; and the *intelligible sense* is responsible for fabulation's translucent intelligibility as well as for the flexibility of its transposition of life's elements.

To make this very long story short, the fabulation that follows on wonderment *intentionally transfigures life's nude facts in accordance with the propensities, virtualities, and factors of the human condition. The result is prototypical models of human character, conduct, societal organization, visions of humanity; the stage is set for the ideals which will fashion a culture's style.*

Taking just the Occidental culture for illustration, we see how the subliminal propensities of the Human Condition have been at work so that the mouth to mouth tradition of folklore has unfolded into formalized art and literature and that the beliefs and rites nurtured in the nucleus of a clan have been amplified in the precepts of formal education. The subliminal propensities corroborate, enrich, amplify, and weed through a culture's tradition, imbuing society with a particular vision. The various cultural styles which thus unfold in the history of nations stem from basic ideals formulated in prehistoric times and which persist through the punctuations of cataclysm, war, defeat and victory, revolution, plague and even cultural epochs. No matter how much the styles of Occidental culture evolve and change, it maintains itself. Ideals going back to Greek, Roman, Medieval, and Renaissance culture still provide the framework for society's institutions and still evoke an echo in the consciousness of the individual.

Honor, perseverance, cunning, beauty, justice, strength, faithfulness, filial piety, these are the virtues and ideals of the Homeric epics, Greek drama and comedy; probity, fairness, and justice are the principles of Roman public life; idealized love, the celebration of life's beauty, admiration for the noble gesture, the quest for the sublime and unearthly, this is our inheritance from the medievals; idealized spiritual beauty, a sense of the dignity of the human being within creation, of self-respect and honor, this is the contribution of the Renaissance to our day. Thus, different sets of ideals have informed Occidental humanity, each unfolding in a progressive corroboration, and each the focus of a distinguishable period.

Although ideals are constantly being nuanced and corroborated and thus evolve into myriads of individual fashions that are then abandoned and either forgotten or revived in novel forms, the ideals themselves persist in the fabric of Occidental civilization. Their presence determines the identity of this culture. The extent of individual education determines how close or remote a person is from the works of literature in which these ideals are set forth. Yet, somewhere within the intelligible aesthetic web that subtends each person's life experience there is surely a chord which would resound with admiration on being confronted with the stories of Achilles' valor, Penelope's fidelity, Antigone's noble devotion to family obligation, Oedipus' pride and absolute commitment to truth, the Nibelungen's powerful magic, Roland's love of country, Le Cid's filial piety and love in the midst of an insoluble conflict, and Dziady's ancestral patriotism.

The course of civilization changes the practical approach to the facts of life and involves constant transformation, ever new rethinking of the great issues that confront the human being, but these great exemplary works of literature are ever picked up anew and revived within the new perspectives engendered by the practical considerations to which the human responds first. This re-thinking, fashionably called "interpretation," consists in reexamining the original "cipher" subjacent to the given ideal. It is to trace an ideal's relevance to the concrete struggles and challenges of life-situations in which it was formed, on the one hand, and to the subliminal propensities of the Human Condition, on the other, and to then refashion it in a way relevant to a new configuration of life-struggles and challenges.

It is relevance to the subliminal propensities chosen for cultivation by a given society that is decisive for the continuity of human inspiration.

At this point let us observe that in these times when the communication between nations, cultures, and individuals is much intensified, we are confronted with trends toward the amalgamation of cultural manifestations. The continuity and identity of each cultural style is now jeopardized. The possibility of a "universal culture," one encouraged by certain ideologies, would require a special investigation which lies beyond the scope of this investigation, our attention being focused here on the onto-poiesis of culture as such, whatever form it may take.

To summarize: culture is what gives continuity to the pulsating life of the individual as well as the human group or society. It is transmitted as inheritance from generation to generation; however, this inheritance is transmitted only when individuals are capable of retrieving ideals and corroborate their meaning with their own living stream of existence. The intentional traces of the past accomplishments in art and literature — ruins, monuments, statues, temples, amphitheaters — which we dig up from the womb of the earth, works of literature recognized to be great over the centuries, manuscripts too that are just discovered, all these are but the dead residual of the past unless we find the key with which to decipher their meaning and make them come alive within our own experiential system.

In literature especially, whatever the genre, human life is lifted above its material soil and finds an adumbration in the spirit. The ideals of happiness, heroism, altruism, patriotism, etc. serve as lamps on far-off horizons. They serve to orient us as we establish institutions for the public good; they help us conduct both societal affairs and our personal lives. The histories of various cultures make manifest how much specific sets of ideals are handed down from generation to generation and provide human groups with identity throughout all temporal vicissitudes and cement their cooperative efforts with a shared spirit.

History also offers us examples of how a people may lose their land, their schools, etc. and may yet maintain unity of spirit — even if dispersed around the globe.

(d) *The Retrieval of Life Significance from a Literary Work*

The end of our hyperbolic survey of the course of cultural evolution lands us where we started from, that is, at our query into how it is that we can retrieve the key to the cipher of an exemplary work. In other terms, we have to shift our attention away from the practical life-meaningfulness of art, especially literature, to the deeply seated ways in which the individual on encountering a work of literature discovers it. We have already belabored the point that it is in the creative vision of the work that the culminating life significance of a masterpiece is to be found. While the attainment of the vision may be seen as the peak of this discovery-deciphering process, undergirding it and orienting it are *the elements and the corresponding elemental passions of the subliminal soul with which we cipher life significance and decipher it.*

As a matter of fact, as we react positively to a literary work, we are involved in a discovery process. A work of art or of literature, at first encounter, strikes us as an unknown. In contrast to the physical objects which belong to the reality in which we dwell in our everydayness, what is depicted on the canvas of a painting surprises us at first sight as a strange universe with the effect of drawing our attention and then either stirring our inquisitiveness or discouraging it. Although it seems to "represent" the views, colors, shapes which we constantly confront in our reality — so much that they become dull reality for us and melt into a general picture surrounding us — it represents reality in its very own way, in such a way that we are prompted to ask "What does it 'represent'?" and then ask "What does it mean?". The closer the representing object approaches what we know in reality, as with, e.g., a good portrait, the more we wonder about it.

An analogous situation is presented by a written text. A book might by its cover already stimulate curiosity about its "content." This content is not directly evident; it has to be wrung from the work's material base and formal structure. Artful covers and titles are meant to intimate the interest the content will have once retrieved. However, whether we have this kind of stimulus or not (we do not, for instance, in the case of an old manuscript newly retrieved from oblivion), just the knowledge that a text can convey a "fable" in some or other form intrigues us. If we are attracted to a text, it is because it promises diversion from everyday life. It lures us away from preoccupations, worries, etc. And yet, in order to wrest from its content its meaningfulness, we have to immerse its moments into the deepest stream of our own existence, our deepest self. Hence, the work radiates into the reality that we deserted for an instant for its sake.

When we embark on the course of discovery following the linguistic rules of the written script or the dramatic sequences of the stage performance, we first seek to retrieve the fabula in which it comes; second, we avidly seek clues relevant to our own personal experience of life.

The story of Don Quixote de la Mancha, or of King Lear, Cyrano de Bergerac, or the Idiot may be read by us principally for secondary reasons (e.g., interest in the history of literature, out of a liking for the atmosphere of a period, or a taste for a particular style, etc.). However, though we may relish the secondary aspects of a work, we still remain on the lookout for "something" that would be the "heart" of the fabula, something that makes its "sense," "meaning," a "message," a "point." In the pursuit of clues to the work we may (and frequently do) mistakenly identify as its central point, the point around which all its material converges and which subtends the script's organization and is intentionally established by its structure as the core of the fabula — a situation therein which is relevant to our own deepest feelings, one that is relevant to our own life-script. We may have a strong need to find in the tale an idea, an experience, a situation that is especially relevant for us. In an "ideal" situation the reader might, due to a noble elevation of mind, respond in his/her innermost to the clues which for him/her make Don Quixote's peripeties, to choose an example, analogous to his/her own and to those of everyone who seeks in matter of fact reality glints of the ideal. Transfiguring this reality for one's own enjoyment does not, however, transform it; we do not even come face to face with the ideal, so deep is the chasm between ideals and facts. The ideal discovery of clues may happen if the reader or spectator's heart is touched at the spot where he has been touched by, say, the cruel ingratitude of his own offspring, or long-term injustice, or the miseries of old age, etc. Such situations abound in exemplary works of literature. The core significance of the fabula incarnated in their scripts has been retrieved over and over again throughout the ages, albeit in differently nuanced responses.

It may, of course, happen that we do not find their central points to be personally relevant and that our discovery of some work only skims the surface then. It also happens often that we find relevant some random moments at which Don Quixote, King Lear, Othello, etc. become alive for us and touch us at our core even though we do not at that point understand them in the essential perspective being established in the script. In the finding of such marginal relevance within a work, the reader may, for instance, see the tragic doom of Oedipus as the result mainly of his excessive pride or of an unconditional commitment to truth, or of the essential blindness of the human being moving about the world understanding nothing about the course of his life. Lastly, we may attribute the intricate web of incidents in which Oedipus unwittingly gets trapped to inexorable *fatum*.

These subjective and personally determined reactions to a literary text account for the "variety of readings" so much discussed today, for the "conflict of interpretations"; they do not, however, legitimize the surmise that the "reading" of the work is arbitrary, that one interpretation is as correct as another. Such a surmise ignores the fact that the core to be retrieved from the script is a fabula. Above all, these options do not justify a presumption of relativity that rules out the correct retrieval of the essential core of a work.

To the contrary, inasmuch as the possibility of relating personally to some or other special point in the complex texture of the work of literature is established by the work's intentional — ideally unchangeable — structure, these clues are intentional *knots of significance* in the pattern that makes the fabula ciphered in the script. They punctuate the significant "nervous system" of this pattern as it emerges from its structural formal/material moments.

CHAPTER FOUR

THE VIBRATION OF ENJOYMENT

(a) *Aesthetic Enjoyment in Its Formal and Material Framework*

Imbibing from exemplary literary works will illustrate best how it is that in a literary medium it is the fabula that ties the knot that ciphers elementary drives, pulsations, strivings, etc. with reference to the subliminal propensities of the Human Condition, laying down a significant sequence to be deciphered by the reader, spectator, listener. The fabula also opens a new approach to literary works, one that takes into account the "methods" of analysis and interpretation applied thus far but which, more, reaches the passions that gives a work its significance, animating it. For the fabula proceeds from the subliminal source, and in following it to that source, we discern the difference between "Nature without" and "Nature within." Then we see how a work's human significance comes alive and makes all the chords of the reader's soul vibrate.

How great writers, playwrights, and poets were clearly aware of this two-fold process that leads, first, from the hidden recesses of the author-creator to the universal significance proposed to the recipient, who, then, in reverse, retrieves it as he brings the work to his own innermost psyche is manifest in the ways in which they devised their work for the stimulation of this retrieval. Though they remain the best judges of this process, making judgments is a role which they share with the public. Critics and philosophers who retrospectively assume the task of analyzing and interpreting the literary work are properly intermediaries between the author and the recipient as they theorize about the rules, devices, structures, etc. that the literary work exhibits and which it should employ in order to perform its role. But this judgement is often obscured by their theorizing and their conventions. More often than not, they misjudge, ignoring or forgetting what it is that a work should do. It is up to the public, the reader, to rectify these errors.

The stage play (drama in all its kinds) is most suited for the illustration of our views because in it the fabula appears in a pluri-dimensional crystallization which means that it by nature only tangentially depends on the script; this prevents the fabula's unwarranted absorption by the

A-T. Tymieniecka (ed.), Analecta Husserliana, Vol. XXVIII, 45–75.

script in analysis or its identification with the "story." Within a play meant for stage performance, fabulation involves our visual, auditory, and intellectual experience. It stands our clearly, and it is not to be identified with any one of them. It comes to its own most clearly, then. It resuscitates the marveling and wonder which initiated it in the first place within the author's creative forge; by appealing directly to our sensorial and intellectual functioning, it sustains enjoyment in the most direct way.

The most striking example of the pluridimensional incarnation of a fabula we find in the Greek drama, tragedy and comedy. The form of the stage play is suitable in every way to the aim of touching and involving the spectator. The Greek convention shows an acute awareness of the play's purpose of getting the spectators out of the current of everyday life, worry, concern, and preoccupations and awakening the spectators' wonderment and arousing their interest and expectations forcefully enough to transport all on the wings of aesthetic enjoyment into a fictional reality radically other than their own, in short, of offering them an intermission in life. This intent is mirrored in the general conception of the theatrical performance and the special construction of the theater. A stage is a tiny fragment of real space excised from the real spatial continuum, and this isolated special locus for the dancers, for the chorus, and for the performance of the play indicates in a striking fashion that a performance is not an event of everyday life and that it is beyond strictly individual concerns, for it is presented in front of the public. Furthermore, the masks used by the actors accentuate the unusual character of the play provoking surprise and the effect of strangeness. Just outward visual appearances already stimulate curiosity and wonderment.

In Aristotle's theory of tragedy, which analyzed and synthesized the principles of the Greek stage play, it is emphasized that the protagonists have to be of a different magnitude than those we meet every day: they have to be either larger or smaller than life. This is brought out to begin with by the *cothurni*, the high thick-soled boots, worn by the actors.

So much for the play's visual impact upon the spectator. The protagonists were meant to personify outstanding characters in every respect. For a tragedy to strike the imagination of the spectator and transport him/her above the pedestrian level of life's petty concerns and to awaken feelings, emotions, and valuations of a higher order, there need be among the protagonists some unusual heroic or elevated characters who do not comply with the pedestrian precept to submit to social conventions but who seek the realization of the most elevated and sublime principles of human conduct that man knows. These characters would never be satisfied with life's ordinary satisfactions but must irresistibly strive for the accomplishment of a moral order suspended upon the sacred. Paying the price of life means nothing to them if by paying it they keep their noble commitments.

If we follow Aristotle's *Poetics* we discover that its seemingly rigid rules and principles were devised in order to give an unfolding drama a close-fitting vertebral column which imparts a dynamic tension to the whole, a tension which grows with gathering momentum. This tension advances from primary reactions, sensory and instinctive, works itself through the network of empirical sensibilities, emotivity, etc., and is concentrated upon the stimulation and working up of the subliminal passions; upon the waves of these latter, the spectator is carried to the heights of his potentialities: he is prompted to unfold the spheres of noble feelings, of virtues, family honor, devoted friendship, fidelity, and respect for the divine order. With the human significance of life thus expanded, empirical existence is dwarfed and appears unworthy of being maintained otherwise.

The chorus plays the role of midwife in Greek drama as it provides the cadence of the rhythm with which the plot unfolds step by step and it knits the specific significance of events. It dwells upon the moments at which this significance is reached, bringing them home to the public. Lastly, it distills the universal significance of the fable in the perspective of all human destiny.

As Aristotle points out, this entire network of the devices intensifies emotions while enlarging the emotional sphere at the ultimate point of emotional discharge, the "catharsis." In the catharsis the entire range of emotions informed by the subliminal significance and universalized to all human experience explodes. It is not in the emotional discharge that is usually emphasized that the "purging" role of catharsis consists, it is rather *in the reaching of subliminal plenitude, an emotional completeness that purifies from all petty life concerns.*

Sophocles' *Oedipus Rex* manifests this intent to progressively work up man's entire endowment and lift the human being toward life-transcending valuations. As in other plays (e.g., *Antigone*), the real events which made the tragic knot that has to be untied lie in the past; their factual significance cannot be changed. However, these facts and their significance are unknown, hidden at the beginning of the action. Had no plague befallen Thebes at the opening of the play, they would have remained in oblivion. When something of such magnitude occurs, however, and its causes in the empirical order are unknown, where else would the king turn but to the Oracle at Delphi? The causes there revealed could then be addressed and the disaster averted.

The action is launched then at the point of a search for reasons that lie in facts past; the plot consists in this search. From the very outset the dramatic — tragic — knot is tied with reference to the oracle, that is, to the highest point of the supranatural: the gods and their moral law for mortals. Thus, the schema of significance to be unfolded by the plot is from the very start suspended on the poles of two orders, the empirical order of facts and the order established for humans by the gods. Human life lies in between and consists, it seems, in attempting to make sense of the interrelations between these two orders. It is by moving between them that Oedipus enters the struggle for understanding the progression of events in which the plot consists. The main line of the plot is Oedipus' burden as he tries all means to unravel the divine oracle; soon he will seek to clarify his own situation with respect to its verdict. Reviewing the turns of his life-course, he believes he can dismiss as implausible any thought that the oracle addresses his own person. However, he discovers facts which demonstrate that he knows nothing about his past; on the one hand, the decisive events of his infancy were hidden from him and, on the other hand, while he had believed that he had in his adulthood acted in full knowledge, he was ignorant of the true significance of his acts: he had not known who the man he was killing was, nor had he known that the woman he was marrying was his natural mother. Lastly, in imposing the penalty of death upon the man whom the oracle singled out, he does not know that it will be himself.

In the progress of this discovery, we are at first animated by a gamut of natural emotions: curiosity, fear, wonder. But then, in steps, we pass to the subliminal moral and sacred passions of yearning for the truth, dread, compassion, renunciation, and expiation as we discover the congenital helplessness of the human being. We follow turn by turn the dynamic heightening of the moral passions until the revelation is completed and all that we hold dear is tied in a tragic knot.

The chorus punctuates the emotional/significant progression by, first, explaining the subliminal significance of events, characters, and feelings and emphasizing the reversal of fortunes in life, thus, drawing out from the brute facts a higher tonality of significance. It further pinpoints the peak moments. Emphasizing the radical reversal of fortune, it chants:

> People of Thebes, behold Oedipus, the King:
> He solved the Sphinx's riddle and ruled this land
> In might and honour; all men envied him
> His wealth and happiness. Now all his might
> Is gone, drowned in a stormy sea of trouble.
>
> (Scene v)

Then the chorus as its last move seeks to stir the deepest subliminal compassion:

> Has anyone ever tasted more unhappiness?
> His wife his mother, His brothers his own sons —
> Can anyone claim more bitter grief?
>
> (Scene iv)

Oedipus' tragedy as unfolded in the plot reveals its universal human significance when viewed in the perspective of its subliminal progression — which is carried by the passions of Elemental Light. Oedipus brings upon himself his doom by following our innermost passion to know, to understand, and to obey what the truth dictates; he is committed to justice and prompted by self-righteous pride. These are the subliminal passions of the elemental light, passions of glory, of life. Yet strangely, in pursuing them, Oedipus slowly moves toward darkness, towards his condemnation. His quest for clarification, lucidity, light leades to the paradoxical discovery of the darkness in which man enacts his life-course. Even as he craves for light, man is condemned to darkness: he does not know what he is about. He is blind as he decides on the acts which determine his destiny. In utmost despair and revolt, Oedipus shouts: "I have seen enough." Darkness without end, darkness pierces his eyes — the very organ of sight — forever.

This desparate act is in its final significance an absolute submission to the unknown decree of the gods and a renunciation of the human prerogative which has revealed itself to be treacherous and delusive. Seemingly the master of nature and earth and priding himself in living in the full light of the senses and intelligence, the human being is, for his not knowing the designs of the gods, blind. Having reached its subliminal peak, the tragedy of King Oedipus, in retrospect, discloses to us its dynamic thread: the passions of light. Swinging from the one extreme of the human situation in nature to the other of the subliminal absolutism of the ideal of the truth and justice, these passions reveal the element of light to be a primordial factor of the Human Condition. This opens for us the study of the correlative elements which will follow in the second part of this work.

The rules by which Greek tragedy conducts and generates subliminal passions as synopsized in Aristotle's theory became the model for Occidental drama. The classical French drama from which we will next draw examples for our study assumed them as absolutes. Nevertheless, there was a striking difference. While the Greek drama used the above-mentioned scenic devices to accentuate the gap between the stage and the everyday world, Renaissance and neo-classical European drama (e.g., Shakespeare, Corneille, and Racine) abandons this convention and brings actors on stage in the costumes of the period in which the play is set, without masks, etc. Moreover, and what is of great interest for the present argument, while Greek tragedy involves the entire gamut of our subliminal passions in their immediate reference to the life-transcending supranatural laws and decrees and pulls us from one extreme to another, the drama of the modern classical period dwells on the intermediary zone of valor, virtue, heroism, glory, honor, revenge, recompense, great deeds, generosity, and noble feelings and actions. And more often than not, love is the center of the tragic knot in this drama. Putting to the side the so-called "sacred tragedies," the plot encompasses the human situation and is closed upon earth and does not extend the subliminal passions to the supranatural. Thus, the ways of arousing and conducting the progression of the subliminal passions, on the one hand, and of amplifying and intensifying the enjoyment of the spectator/reader/listener, on the other, are calculated differently. Fabulation assumes a different form.

First of all, with the chorus eliminated, the explication of and deliberation on the nature and nuances of feelings, their transformations and increasing or declining belong to the fabulation itself, belong to its intimate mode of unfolding; so does the discovery of truth through a series of events as well as witnesses.

Secondly, the theater techniques used are limited to the play on the stage — its performance; understanding of the play's significance inheres to the fabulation. We shall see in our subsequent discussion what happens if a technique used in the structure of the play comes into conflict with the intent of the fabula.

Third, when Renaissance courts commissioned works of art, the profane passions came to the fore and sacred dread and pity were replaced at the peak of a work's subliminal unfolding, which no longer pierced the transnatural spheres but knew only the earthly passion of the sublime; this passion pertains to the generosity and nobility of the human heart and its capacity for disinterested self-sacrifice rather than to the transcending passions of light.

Thus, at the heart of the plot lie conflicts of love and hatred, jealousy, vengefulness, and forgiveness. As mentioned above, the disentanglement of the conundrums of attraction, feeling, desire gives occasion for the elaboration of the great passions of love with its sublime elevations and plunges, and this elaboration belongs to the fabulating function which constitutes the core of the plot.

Fourthly, it is most noteworthy that because the drama is definitively placed within the fabulation as it focuses upon the game of emotions and subliminal passions, the passions are clarified by means of monologues, dialogues, and confidences given, all of which are specifically designed for this purpose.

(b) *The Rising Tide of the Heart*

With all this in mind, where could we find more appropriate examples than in the classical French drama? It was Aristotle's *Poetics* that gave the principles and rules for tragedy that were considered by critics and learned literary people to be absolutely valid. In both of the most representative dramas of this period, *Le Cid* of Corneille and *Andromaque* of Racine, this established convention is maintained. However, together with the above-mentioned four points detailing its inner transformation, some leeway is provided the playwright — and both Corneille and Racine take advantage of it, albeit in different ways. As a result, their examples sharpen our argument. Indeed, they seem even to have fully assumed the intent we propose for drama.

In *Le Cid* a conflict between filial piety, family honor, and elevated amorous passion is brought to its highest peak in the sublime feelings of Don Rodrigue who performs great deeds for his country to offer atonement for a duel and disobedience to the king and who seeks only the intimate forgiveness of Chimène for his having killed her father even to the point of renouncing happiness with her to obtain it. His sublime example leads to a subsequent outbreak of elevated generosity and general reconciliation and a happy ending, which transforms the tragedy into a "tragi-comedy." To carry out this design, Corneille does not pay meticulous attention to the classical rule of the "three unities." We witness in the unfolding of the fabula which begins on a high note as the two lovers consider fulfilling their great dream and being united with the consents of their respective fathers how the play of emotions unfolds in dialogues during which feelings and their expression lead to ominous gestures (such as the slap at Don Diégue by Chiméne's father, Don Fernand). We witness how the deliberations on feelings and the mounting of passions occurs in intimate and passionate dialogue between Don Diégue and his son, the latter and Chimène, Don Fernand and Chimène, as well as in the confidential exchanges between Chimène and Elvire, her confidante, and between L'Infante and her confidante, Leonor.

The public already at the first performance was carried away by this emphasis on the nuances of unfolding subliminal passions. Its enjoyment of the noble spirit of the play and especially of the generous disinterestedness of Don Rodrigue had no bounds. However, the audience's enthusiasm found striking opposition from the side of learned academicians and critics. This is not the place to enter into the famous "querelle de Cide," but this controversy confirms our point. The erudite academicians reproached the play for its violation of the sacrosanct structural rules: first, for having combined comedy with tragedy by the happy denouement; second, one of the critics, Georges de Scudery, attacked the play for lacking diversity in the unfolding of the plot, an intrigue;[11] further, it was objected that there are no secrets to be revealed; and, lastly, there is no dramatic knot to be untied or cut. In point of fact, these reproaches might be justified. Certainly there is only one logical movement in the plot from the beginning of the action to its fulfillment. It is also correct that the plot disregards an absolute adherence to the rule of the "three unities." However, and here is the point, with the displacement of the focus from the deeds of the plot in their significance to the intimately personal sphere of experience, then, quite naturally, the rigid rules of structure gave way somewhat to different means for stirring the deepest emotions and bringing the passional sphere of the subliminal soul to its highest level and densest intensity. Here, it is by means of the inner advance of the emotional sphere in its transformation that the strokes that punctuate the next steps are hit. It is by means of the *inner conflict* in the psyche and soul of the protagonists that the plot proceeds instead of by means of discoveries of truth or, as is the case in *Antigone*, in the peripeties of a heroic pursuit.

The audience is seized by the unfolding passions in this play of contrary moves of love and hatred as filial piety calling for noble vengeance opposes adoring devotion to a beloved woman. The inner conflict of Don Rodrigue illustrates best the intensified subliminal passion:

> Père, maîtresse, honneur, amour,
> Noble et dure contrainte, aimable tyrannie,
> Tous mes plaisirs sont morts, ou ma gloire ternie.
> L'un me rend malheureux l'autre indigne du jour . . . .
>
> (Act I, Scene vi)

Two noble passions are clearly opposed, and either choice of action has equally contrary and fatal effects. The insoluble tragic knot is tied. Happiness is impossible without honor; and what is life if happiness is lost? Hence, Rodrigue lets loose his emotions. Now he would revive his deepest hopes for happiness, now he would avenge his house; now he feels the slap his father received, and now, the unbearable prospect of losing his love:

> Cher et cruel espoir d'une âme généreuse,
> Mais ensemble amoureuse,
> Digne ennemi de mon plus grand bonheur,
> Fer qui causes ma peine,
> M'es — tu donné pour venger mon honneur?
> M'es — tu donné pour perdre ma Chimène?

Having thus placed all in the balance and seeing no escape from the conflict, he concludes with a reaffirmation of both of his deepest commitments, before which death seems the only solution:

> Il vaut mieux courir au trépas.
> Je dois à ma maîtresse aussi bien qu'à mon père:
> J'attire ses mepris en ne me vengeant pas
>
> (Act I, Scene vi)

Stirred by Don Rodrigue's adherence to filial devotion and the defense of family honor heedless of the cost, on the one hand, and by his devotion to his beloved, on the other, and then struck by the deadly blow by which Don Rodrigue carries out his filial duty, the deed by which he foreits his happiness, our emotional register is aflame and our enjoyment gives way to subliminal elevation as we participate in both sides of the conflict tearing Don Rodrigue apart. We are thus raised to the elevated sphere of the significance which facts and deed may acquire for man.

We participate, too, in the simultaneous play of the emotions that tear apart Chimène. She confides to Elvire (Act III, Scene iii):

> Dans un malheur si grand to parles de repos.
> Par où sera jamais ma douleur apaisée,
> Si je ne puis hair la main qui l'a causée?
> Et que dois-je espérer qu'un tourment éternel,
> Si je poursuis un crime, aimant le criminel?

From this synthetic statement of the tragic conflict into which she is plunged, Chimène moves to unfurl the complex of passions which carry it:

> S'est peu dire aimer: je l'adore;
> ma passion s'oppose à mon ressentiment;
> Dedans mon ennemi je trouve mon amant;
> Et je sens qu'en dépit de toute ma colère,
> Rodrigue dans mon coeur combat encore mon père.

There follows a minute description of her feelings.

It is obvious even from these briefest excerpts that the tonality of the passions involved in these conflicts is already subliminal. They sound in the sphere definitively above primitive self-interest and beyond the orbit of crude survival concerns and the accompanying hostile attitudes and feelings toward anyone who contradicts them and hinders the accomplishment of aims. These conflicts, both sides of which strike the sublime chords of the soul, have their subliminal core fully developed in both protagonists — and in the Infanta to boot — namely, the basic motivation of *generosity* of heart. In each of the knots of conflict a magnanimous recognition of the plausibility of each side's reasons for standing its ground comes to the fore. This comes out particularly in the dialogue between Don Rodrigue and Chimène which follows the avowal quoted above.

> Ah. Rodrigue, il est vrai, quoique ennemi,
> Je ne puis te blâmer d'avoir fui l'infamie;
> . . . Je ne t'accuse point, je pleure mes malheurs,
> . . . Ma générosité doit répondre à la tienne;
> Tu t'es, en m'offensant, montré digne de moi;
> Je me dois par ta mort, montrer digne de toi.
>
> (Act III, Scene iv)

After being moved by the heart of Chimène who maintains her love but keeps herself from forgiving Rodrigue, we are moved to the heights of the sublime by Don Rodrigue's readiness to give his life to expiate for the death of Chimène's father — for the sake of obtaining Chimène's forgiveness and nothing more.

These conflicts find resonance within us, and we naturally enter into the deliberations of the protagonists as revealed in their dialogues, the deliberations which exfoliate their feelings, emotions, tendencies and which constitute *the field of the soul's struggle to release its subliminal virtualities.* Following them moves our own soul until the sublime is our own sphere. Corneille's *Le Cid* opens for us the hidden springs of the subliminal soul as such.

We may say that this illustrates Descartes views. In his treatise on the passions, Descartes asks himself whether all the human stirrings and emotional states are fruits of the human natural endowment, that is, physiologically rooted, or whether the soul may have some "passion" of her own. He explicates generosity as an intimate passion of the soul which it releases from its own resources and from which flow all other elevated feelings, emotions, etc. that cannot be attributed to an empirical origin.

We would corroborate what Descartes says about "generosity" by noting that here is a passional complex proceeding from a subliminal progress that is already far advanced; that is to say, his view assumes that the human being has already brought himself to a moral level at which he respects himself to the point of not surrendering to any low or demeaning egotistic impulses, but is capable of recognizing within the other what is due to him in his own right — and of striking a balance between what is due to oneself and to the other. Indeed, the passion of generosity so basically conceived reveals itself in great classical theater to the fullest.

(c) *The Swing of the Pendulum*

Moving now to *Andromaque*, the great tragedy of Corneille's successor, Jean Racine, we will bring our argument to its culmination. Racine's *Andromaque* is, indeed, particularly instructive for our purposes. Racine must have not only known and studied in detail all the ins and outs of the "querelle de Cide," but also meditated on the issues involved, for he gives them his own solution in his work. The central issue is that of the relationship between the prerogatives of formal perfection versus those of the emotional content expressed. (Racine not only studied but also translated Aristotle's *Poetics* giving it minute annotation.) We may see in *Andromaque* a full-fledged example of his own stand on the general issue of what is the significance of drama and, consequently, on how a tragedy as a stage play should be conceived.

Most interestingly, *Andromaque* may be considered to be a "sea drama." The drama, in fact, begins and the tragic knot begins to tighten when Orestes, as an emissary of the Greek cities to Pyrrhus, King of Epirus, throws his anchor on the shores of Epirus creating a critical predicament for Andromaque, the widow of Hector and the mother of his son, Astyanax, both of whom are held in the palace of Pyrrhus. The drama unfolds, then, on the shore. And then, when all the forces of heaven and hell, angelic and demonic are lifted to the highest state of expectation, and then all lofty plans are brought to the ground in an explosion, and the play reaches its conclusion, Orestes, falling into a demented rage, has to be carried to his ships by his companions and depart. In fact, it is the background of the changeable sea which gives this drama, from Orestes' arrival to the raising of the sails for his departure, a particularly poignant, self-enclosed unity.

It is the sea which brings people together and puts them asunder, thus giving an occasion for linking people's destinies. Even before the plot unfolds, the four protagonists who figure in its three dramatic conflicts — Andromaque and Pyrrhus, Pyrrhus and Hermione, Hermione and Orestes — are drawn together by voyages across the sea. And at the dramatic point when the turmoil of emotions and desires is greatest and and at odds and when the entire course of events could have been far different, disaster is averted because the sea offers escape. Then

Pylades urges Orestes to cut the tragic knot by capturing Hermione and carrying her away to Greece, his ships being ready and the sea at high tide, battering the walls of the city, and the night being propitious for carrying his beloved "prey" to his vessels (Act III, Scene ii).

In *Andromaque*'s first public performance in 1680 (it can be assumed according to the stage directions of Racine himself), the action at Pyrrhus' palace took place before a range of white ancient columns that stood against the background of a sea covered with white vessels.[12]

The sea offered open space for escape, freedom from the insoluble knots of the situation on land. But what is it that drew people across its vast, and in past ages, uncharted and unexplored distances at the mercy of unpredictable winds and the mischief or benevolence of the gods, through perilous adversities which do not spare even intrepid navigators? What prompted and prompts them to dare the voyage, to light-heartedly risk their lives? What do they pursue that makes them brave the storms and sirens?

After the fall of Troy, where the hand of Pyrrhus felled Hector, there has come to Buthrode by a network of different sea routes Hermione, the betrothed of Pyrrhus; Andromaque, his captive; and lastly, Orestes, the emissary of the Greeks. The routes they took over the waves were different, but the motor and the compass of this fateful encounter, the causes of its threefold dramatic knot were violent or inwardly gnawing passions of love, each different in kind. There is the sensuous, tender love of Hermione, which is ready to forgive Pyrrhus for his enchantment with and elevated devotion to Andromaque — a love which knows the gamut of human feelings, from the sensuous infatuation which inundates his being through the range of all worshipful and moral sentiment, and which is crowned by the most immense generosity of heart. These two loves alone give us two great variations on the theme of love. However, Racine brings forth another love, the enduring, irresistible, deep-seated, gnawing love of Orestes for Hermione, which though unrequited seems to endure all, which feeds itself on its very substance and is revived by flashes of vain hope. The center, however, is occupied by Andromaque and her absolute, faithful dedication to the memory of her lost husband, a love which she revives in her maternal devotion to their son. Each of these dominant passional complexes is accompanied by an assortment of virtuous sentiments — self-respect, honor, the cult of the family and the ancestral past, etc. Racine deploys in *Andromaque* a compelling panorama of human feeling, and he molds the form for the sake of bringing them in their most complete significance to life within the spectator, reader, auditor.

He most appropriately projected this fabula against the element of the sea. The passions of the heart, the innermost personal sphere of the human being, whether petty or sublime, heroic or cowardly, benevolent or malefic, reveal themselves at the surface of life against the background of the powerful element of the sea.

In point of fact, the quintessence of his conception of tragedy is that it should involve simple action with only a small amount of stuff (*matière*) so that it may unfold in one day and which, "advancing by degrees towards its end," "*is sustained only by the interests, feelings, and passions* of the characters." He concludes against those who are principally concerned as to whether the formal rules of tragedy are respected: "One does not write for pedantic critics," he says, but for "the small number of wise people whom one attempts to please." (First Preface to *Britannicus*, 1670) Rules are important, but "the principal rule is to please and to move." (Preface to *Bérénice*, 1671) All the other rules are made only to achieve this primordial goal. He shows what writing is all about. The major quality of a writer is for him "the gift to please." (Preface to *The Thébaïde*, 1664)

Of course, the question is what it is that he understands by "moving" the public or "pleasing" it. That he is at the furthest possible point from arousing the reader or spectator to vulgar gratification of any kind, to any pedestrian, trivial, everyday response is obvious from the fact that he conceives of the feelings and the passions that are meant to be "moved" within the formal precepts of antiquity and according to the refinement of the mores of his times. How the audience is moved and what sort of "pleasing" is sought will come out in our inquiry into the drama itself.

First, he accepts the great formal rules of the Greek drama and maintains carefully the "three unities." However, in an innovation, he gives decided priority to the fabula with its emotional/passional core. Second, the plot is thus carried by the inner conflicts of heart of the *dramatis personae*, which their actions then reflect. Lastly, although he adheres to the Aristotelian principle that the protagonists should be "greater or smaller than life" and accentuates their distance from everyday existence by choosing them from foreign cultures and the historical past, he, nevertheless, modernizes the drama further by combining the principles of those places and times with the most refined standards of the high culture of seventeenth century France, styling accordingly their experience and their expression.

By making use of the classical devices while endowing them with a primarily psychological dynamism, he renewed classical drama and yet opened the way for the subsequent relaxation of its rules and then its destruction or completely arbitrary transformation. It is precisely the masterly way in which Racine uses all the classical formal devices and combines them with his own incomparably skillful and imaginative fabulation that allows him in *Andromaque* to make the whole sweep of the passional circuits of the human being vibrate. In order to strike the proper chords in the viewer to the full, he applies a regulative schema to give the play its dynamism. The dynamic progress advances by leaping between contrary tendencies of the heart in the inner struggles of each of the four characters whose lives are dramatically entangled. This play of contrary passions cannot fail to shake us. This progression is punctuated by decisions taken at the end of each act which bring together the varied lines of emotional deliberation and passional striving; it makes clear the state of mind of the protagonist and builds expectation. And then a minor scene follows each of the violent outbursts of passion, exasperation, or distress, which gives pause to the precipitation of conflict. This allows the spectator to take in the gravity of the situation, to ponder it in his own heart; briefly, it allows a meditative interlude which secures the mood of "majestic sadness" (*tristesse majestueuse*) — seen by Racine as the specific climate of tragedy — to grow in the midst of a succession of turbulent emotions.

However, here is a minute display of rising passions through innumerable sentient and emotive changes in each of the characters involved in the making of this tragic knot, with each of them pulling one of the strings that carries the drama. The register of feelings is not only far vaster than in Corneille's *Le Cid*, but it is subtended by the lines of the four modalities of love distinguished above, and it seems to cover the entire repertory of the passions of the heart. It is its inner dynamic orchestration that particularly attracts our attention.

We are overpowered by the play of passions, which veer, first, in one direction advancing through all the gradations of feeling to the peak of love or hate and, then, when our sentient system is completely engaged, suddenly reverse direction, throwing us from the saddle as the emotional stream sweeps to another peak.

Pyrrhus persists in an unrequited love for Andromaque in the hope of winning her yet. Now, he is put on the spot by Orestes' arriving to ask him to deliver her young son Astyanax over to the revengeful Greeks who want to put him to death. He summons all his forces and strikes all the chords to make a last attempt to have her. He will seize on a situation that seems propitious. In an incomparably controlled and balanced but frenetically passionate way, he adjures Andromaque to accept a compromise:

> Au nom de votre fils, cessons de nous haïr.
> A le sauver enfin c'est moi qui vous convie.
>
> (Act III, Scene vii)

He would convince her to overcome her hatred of him for killing her husband in battle and her oath of eternal faithfulness to Hector by asking her to magnaminously accept him for the sake of her son, to whom she is as deeply devoted as to his father. Then, changing his approach, he displays the tender devotion he feels for her. To bring her to see its significance, to enamour her of it, he entreats her:

> Faut'il que mes soupirs vous demandent sa vie?
> Faut-il qu'en sa faveur j'embrasse vos genoux?

We are touched by this voice coming from the very depths of his soul. We are ready to give this voice our confidence. Then, with a resolute decisiveness that shows how the will seizes the reins of his feelings, he proposes the scheme: he will lead her to the temple where in a ceremony prepared for his nuptials with Hermione, it is Andromaque whom he will wed, and crowning her queen, he will in true paternal fashion take Astyanax under his wing, protecting him and supporting him in life. Pleading with her and justifying the plan to her, he changes tone again to deliver a foreboding message. This is not another way of wooing her, it is an ultimatum:

> Mais ce n'est plus, Madame, un offre à dédaigner:
> Je vous le dis, il faut ou périr ou régner.

With one alternative being death, only terror is inspired. But then Pyrrhus strikes a lovelorn note as he expands on his torment and his frenetic suspense, his ever disappointed and ever renewed élan:

> C'est craindre, ménacer, et gémir trop longtemps.
> Je meurs si je vous perds; mais je meurs si j'attends.

What a range of emotions are synthesized in this picture of his soul torn apart! Could an appeal to a woman's heart play on more tender strings? Striking them as an accompaniment to his main theme, that is, of his promised protection and devotion to Astyanax, he asks her to bring her son to the temple.

Here, in radical contrast to his now violent, now tender, no magnanimous devoted passion, he makes a threat and presents a menace as radical as that devotion, one which seizes us with horror, as he concludes:

> Et là vous me verrez, soumis ou furieux,
> Vous couronner, Madame, ou le perdre à vos yeux.

All in all, it seems that Pyrrhus in his proposal strikes a balance between that magnaminity that surges out of the emotional turmoil which torments him and the reasons Andromaque steadfastly alleges for her spurning him — by summoning her to a like magnaminity, that of a mother ready to save the life of her son at any cost.

Alas, in vain. Critics of the seventeenth century accused Andromaque of being "une coquette verteuse" who teases Pyrrhus and evades him. There is a deeper reason why she does not forgive, reconcile herself to her fate, and in this way reach a noble elevation of spirit. She is not the type of the romantic lover (*une amante*); her role is that of the widow absolutely determined not to betray her husband and of the mother who sees her husband in her son and is ready to take all risks, to scheme and deceive, for his sake. Maternal love is here the virtue to which all others, even magnaminity, have to submit.

Why does she show herself deaf to Pyrrhus' amorous entreaties as well as to his noble stance? In her own way she draws us into another perspective when in the text scene she passionately reveals her own innermost struggle. In an exchange with her confidante, Céphise, who tries to make her see how far Pyrrhus is going in his generosity, to the point of forgetting all that he owes to his country and to his forefathers in order to make his enemy's son his heir and of raising her, his captive, to the rank of queen restoring to her her ancestral dignity and securing her son's future, Céphise argues that Hector himself would favor her changing her heart toward Pyrrhus. Yet this scene, which promises us a pause after the vehemence of the previous scene and the crucial decision made in it, brings us to another torrent of emotions as Andromaque in response delivers herself in a rapid crescendo of all her hatred. Andromaque, in fact, is revolted at the idea that Pyrrhus could ignore the claims of the past. With mounting vehemence, she evokes sensuous images of the carnage at the walls of Troy, of her father and two brothers lying in their own blood, of the horror of the flames and the screams of the dying and the yells of the conquerors. Lastly, there is the memory of Pyrrhus, covered with blood and eyes on fire, making his way through Troy's burning palaces. Of the picture within which she sees him, nothing may change.

But when Céphise concludes that there is nothing else, but to see Astyanax die, there is again an instantaneous reversal of direction. Andromaque turns from her consuming hatred to the vivid memory of the tender and moving farewell of her adored Hector and of his entrusting Astyanax to her as the token of their love with the words:

> Si d'un heureux hymen la mémoire t'est chère,
> Montre au fils à quel point tu chérissais le père.

Now flooded with revived love and tenderness, she balances her hate for the conquerer with her love for the son. The latter wins, and she decides to accept Pyrrhus' offer. It is transparent, however, that blinded by raging hate and by fear for her son, she does not clearly see how elevated Pyrrhus' attitude is. She asks:

> Crois-tu que dans son coeur il a juré sa mort?
> L'amour peut-il si loin pousser sa barbarie?

Having decided to save the life of her son, she now leaves in exasperated turmoil to commune with the spirit of her husband and learn his will. We know the result of her discernments.

In her exchange with Céphise (Act IV, Scene i), she assumes a heroic, composed attitude, explaining to her confidante her "innocent" strategy. She will save her son, but, faithful to the memory of the dead and to the ardor of her farewell with Hector, she will not fulfill her commitment to Pyrrhus completely. She will take her life rather than betray her husband. In the quietly fervent, suffering tonality of her emotions, a poise beneath which we feel a supreme effort to overcome tumultuous emotions, there is a heroic pathos. This mood solemnly deepens as she gives Céphise her last recommendations on how Astyanax should be educated. She — at last and on the wings of maternal love — lifts herself above the crudeness of hate, mistrust, vengefulness, etc. in recommending that her son should be admonished to have no thought of avenging his father and his forefathers. On the contrary, he should recognize what is due Pyrrhus:

> Nous lui laissons un maître, il le doit ménager . . .
> Il est sang d'Hector, mais il en est le reste;
> Et pour ce reste enfin j'ai moi-même, en un jour
> Sacrifié mon sang, ma haîne et mon amour.

Standing between Pyrrhus and Andromaque, protagonists with rival perspectives, who even when their passions peak in exasperation still uphold the supremacy of reason and balance the innumerable emotions at odds and maintain a posture of dignity, is Hermione, a young girl who has no consideration other than to regain the affection of Pyrrhus, the betrothed who neglects to follow through on his commitment to her and who seeks release so that he can take Andromaque as his queen. When Hermione speaks, she maintains the tone of an imperious princess. But she is love forlorn and tossed from hatred to love and then back as now fury is intensified and now tenderness. We cannot fail to be moved by precisely the "natural" lack of restraint in her outbursts of feeling. Already at her first appearance we see the young girl of any and all times swept away by conflicts of love, betrayal, shame, and humiliation — feelings which make her vacillate between tender love and fierce hatred, hope for happiness and desire for revenge. We hear her confide to Cléone:

> Si je le hais, Cléone! Il y va de ma gloire,
> Après tant de bontés dont il perd la mémoire,
> Lui qui me fut si cher, et qui m'a pu trahir,
> Ah! Je l'ai trop aimé pour ne point la haïr.

As she shares her feelings with her confidante, who attempts to reason with her using arguments of cruel lucidity, we see her juvenile weakness and her helplessness in the precarious position in which she finds herself and are touched. Moreover, as she takes us on her violently shifting course, we are brought to discover the unfathomable depth of great passions, a depth which cannot be explained or understood. We are led into the abysses of the human heart when Hermione, after numerous swings of the pendulum, in a paroxysm of disappointment and furor decides to resolve her sad situation by using Orestes, who unconditionally adores her, to kill Pyrrhus. (Orestes has gladly come to Epirus to take Astyanax because the mission allows him to see

Hermione.) This decision is made after Pyrrhus had come to her first to finally make their wedding plans — which sent her into lyrical amorous ecstasy — and then had returned seeking to be released from the engagement. The pendulum swings again and again. First, Hermione urges Orestes on to kill Pyrrhus in the temple, and then, after the deed is done, she reacts with incredulous rage and roars at him for daring to have actually done it. In this reaction there is opened for us a secret tunnel. We see now beneath the surface of Hermione's previous displays of passion. The swing of the pendulum goes deeper still. Her reaction of stupor and then of furor reveals what could not have been comprehended. Hermione, heedless that it was she who instigated the fatal deed, shrieks in horror:

> Barbare, qu'a tu fais?
> De ton sort qui t'a rendu l'arbitre?
>
> * * *
>
> Ah! Fallait-il croire une amante insensée?
> Ne devais-tu pas lire au fond de ma pensée?
> Et ne voyais-tu pas dans mon comportement
> Que mon coeur démentait ma bouche à tous moments?
>
> (Act V, Scene iii)

How can one guess the true wishes of a woman in love? "Le coeur a des raisons que la raison ne connait pas." Hermione illustrates perfectly Pascal's dictum; more, she makes us realize *the unfathomable distance between the two extreme movements of love and the innumerable movements of emotion, feeling, passion that the pendulum swinging from one to the other has to traverse.*

Orestes now cries out to the gods as the masters of fate. He accuses them of having with premeditation crowned his unfortunate pursuit of love with his committing a murder that brings him not the hoped for reward but his beloved's hatred. He sinks into demented horror. Thus closes this fabulation's symphony of the passions of the heart. In the rainbow of significant sentiments that rises from the opaqueness of Nature and which the human being projects before he returns to its womb, there is a sphere of existence specifically his own. Great literature evokes for us this sphere making it even more our own. The greatest periods of culture are crowned by literature that display this rainbow.

(d) *Beauty and/or the Beast*

We have seen how the classical theater nourished the culture of its times with the most rarefied fruits of creative imagination and how the elevated spirit can infuse human life with significance. But then we can see the literature of a succeeding period swing to the opposite pole and draw only from the opaque depths of nature where light does not penetrate and rainbows are unknown. Indeed, the contemporary emphasis on the "dark forces" in the human being is nothing new.

This trend in popular and even serious literature should not be attributed to merely the influence of pragmatic pursuits which extinguish "disinterested" aspirations and to subsequent forgetfulness of the higher aspirations of the subliminal soul – heroism, the moral virtues, generosity of heart, honor, filial piety, etc. Already in the nineteenth century the naturalistic tendency in great literature claimed to retrieve "the whole of life," in Zola's words, and in order to vindicate the reverse of the coin that classical and romantic literature favored, it made primary the unrelieved violence of human nature on its merely vital plane.

Here an essential line has to be drawn between what we here understand by "passion" and what it means in the naturalistic perspective. In point of fact, the focus on passion, passional intensity, pitch, fervor, violent emotion, etc. may easily lead to grave misunderstandings. We may mistake the *subliminal* passions of the soul with vehement reactions, frenzies, rages, psychological upheaval of all sorts, violent outbursts, and the unrestrained turmoil of indomitable instinctual drives. In all literature of the highest spiritual level we encounter contrasting personalities: generous and noble souls who stand in contrast to villains and petty figures who are often strangers to any subliminal emotion and, who even if tempted to follow a decent and honorable impulse, are incapable of resisting the subterranean gnawing forces of hate, envy, vengefulness, irrepressible impulses to hurt, to humiliate, to upset, to kill, or who are simply too weak and too passive to move in a direction which calls them away from their usual moral indolence. But naturalistic literature onesidedly focuses on this "dark" side of humanness, seeing in it its true manifestation and dismissing the creative uplifting of man into culture.

The naturalistic tendency is to uncover and bring to the fore the vital, empirical maze of feeling, and this leads to a transformation of the inner rules of writing. As we know, Strindberg adopted, in his naturalistic period, some of the rules that brought about the radical transformation and corrosion of modern classical drama, and of writing in general. First, in his Introduction to *Miss Julie*, he says to those who go to the theater "clamoring for the joy of life" and who criticize his tragedy for being too sad that he finds the joy of life "in its strong and cruel struggles."[13]

We may see in this declaration a statement of the first rule of naturalistic literature. Moreover, he points out, and this directly in opposition to Racine's precept that "the action should be simple," that "in real life an action — this, by the way, is a somewhat new discovery — is generally caused by a whole series of motives, more or less fundamental . . . ." This corresponds perfectly with the move away from the heightening of emotions to reveal a subliminal passion and toward a reaching down into the preconscious maze of incoherent turbulent impulses, reactions, etc. which *in their disorder* motivate action. *Miss Julie* is a case in point. Strindberg, believing that he is "up-to-date," enumerates his reasons for the tragic fate of his character. He emphasizes that the "Naturalist . . . knows the richness of the soul complex" — the soul being identified here with the person's physiological and psychological hereditary and cultural circuits — and "realizes that vice has a reverse side very much like virtue."[14]

But the fascinating point he makes about his characters has to do with culture: "My souls (characters) are conglomerations of past and present stages of civilisation,"[15] and after an enumeration of the other bits and pieces from which they are made, he most significantly reveals to us the entire mechanism of Miss Julie's character: "Miss Julie is also a relic of the old warrior nobility now given way to the new nobility of nerve and brain." She is a victim of many precedents which Strindberg attributes either to the family past, or to the "fate or Universal Law of former days."[16] He writes: "The Naturalist has abolished guilt with God, but the consequences of the action — punishment, imprisonment or the fear of it — he cannot abolish." Obviously, the Universal Law of the previous period of culture *is* abolished, but its deep-seated traces remain. Miss Julie is no longer inspired by faith in honor and noble feelings; she does not feel impelled to virtue and duty. And yet she has either inherited or acquired "the sense of honor." Strindberg attributes this inheritance to the continuity of the upper class, which still upholds some remnants of the ideals of its forebears.

The heroine of this play is, in fact, caught at the borderline between two radically different world-views: she is used to coarse behavior and to the modern turn of her mother's mind, and this leads her to succumb easily to her primitive instincts. But then, in a reversal, instead of proceeding with the same attitude, she is tormented by a remnant of the feeling of honor and feels guilty for having demeaned herself by falling into the arms of a socially inferior man. Her inner predicament has the overtones of the modern conflict of the social classes instead of those of a truly moral conflict. And yet in the confusion of her feelings, she is not capable of rising to any constructive solution for her predicament, neither the pedestrian-practical one suggested by her lover, nor a noble gesture or attitude of any kind. In her inner struggle we witness a weak, culturally degenerate being who is neither here nor there. When she succumbs to her weakness, her self-destruction is not an act of honor. There could not have been a more obvious lesson to be drawn from Strindberg's drama than this: ideals, sentiments, and moral principles determine a culture. They inspire and imbue life and its predicaments with sense.

Strindberg's naturalism still moved at the border between the cult of the highest ideals, great deeds, and of models who incorporate them and the modernity which, as he pointed out in the quote above, values individual force and acumen. With Zola (who admired Strindberg's dramas greatly), we move, deliberately, into the realm of the life of the common man. We also enter into the basement, and sub-basement, of human instincts, of primitive strivings and drives which have barely known the touch of the creative orchestration that humanizes them. These forces have acquired only the compass of life-subservient, calculating reason and have not entered into the transforming subliminal forge. In Zola's serial novels, *Les Rougon-Macquart, Histoire naturelle et sociale d'une famille sous le Second Empire*, an epic of human life propagating itself through generations, people are viewed precisely in the "naturalistic" perspective, that is, as they function on this basement and sub-basement level. "Naturalism" has the merit of showing us this dark invisible stream of half-blind forces which are only half-informed by specifically human moral and aesthetic significance; they stand informed just enough to effect in the characters the shift from animal innocence to human moral brutishness. In the novels, *L'assomoire, Nana, La bête humaine*, we are caught in the turmoil of brute drives that lead unavoidably to outbreaks of violence. This reveals to us what human sentient and emotional existence is doomed to become if our instinctual strivings are processed through the filter of the basic functioning that establishes the human condition within the animal realm but are not raised to the level of the subliminal soul with its specifically human proficiencies. While the violence, the terror, the vehemence, the urgent strivings, the vigorous drives, the disorderly turmoil, etc. which we witness are informed by the vital significance of life and are partly capable of using organic, vital, and psychological forces to serve the interests of survival, yet these forces being untamed in their vigor have no power to restrain the "counterforce" of malefic strivings. They retain animal vigor but have lost the moral innocence of the beast.

Zola's *La bête humaine* provides a striking example of existence shaken constantly out of healthy life-routine by the outbreak of irrepressible violence erupting from misdirected moral reactions which lay ready to pounce from the sub-basement level of primitive functioning. The possessive jealously of Robaud — the counterpart of desire, not love — over his wife erupts in an instant; no moral or aesthetic impulse allows for an instant of reflection. An instinctual current rages overwhelming the entire psyche which throws itself, first, into brutality directed at Sévérine and, then, into the execution of a frantically schemed act of vengeance. We are shaken with dread and lose our usual composure as the balanced surface of life's business opens to reveal a subterranean labyrinth of confused primal circuits. And there is no release from this labyrinth. Robaud, indeed, is not "freed" from the horrible shock and disappointment he felt on learning of the relations between his wife and her patron. The disgust which inundated him, and rightly so, when he realized that his felicity was based on a fraud is not dissipated by the death of the culprit. Instead of finding the relief for his pain, he finds that the deed completely upsets the balance of his life. Having lost the compass by which he had steered his life in the past, he sinks into reflex habit. In his inner confusion, he seeks to forget and to distract himself, but he is no longer capable of maintaining regular interest in everyday routine. The reader, with a shrunken heart, settles into the subterranean existence of so impoverished a life, a stygean world any exit from which seems hopelessly blocked.

Zola probes even deeper into the abysmal opaqueness of morally unharnessed and aesthetically unmolded drives in the character of Jacques, the subsequent lover of Sévérine, a man driven by an erotic desire mixed with the desire to kill — a drive which surfaces without restraint and with blinding violence within his primal, primitive circuit of functioning. Once the tentacles of this urge have hold of him, there is no exit; he does not free himself from it. Incapable of pulling himself up to the subliminal level, Jacques, in contrast to Robaud, remains in a regular, placid, and self-controlled frame of mind: a brute of a different kind. Between them is Sévérine, a non-violent beast who leads a lucidly calculated, narrowly framed life of instinct, sensual pleasure, and trivial joy; she submits to necessities and swims contently on their surface. A non-violent beast.

We leave the scene of this supposedly "natural" life in abject depression and full of questions. Is that all that there is to a life? Is it possible in someone's life that there can be no salvaging device for one's hurt, no reprieve from one's incalculable folly in following one's instincts? Does the inner reality of man preclude a second chance? Is such a life of violent wild urges, like those of Jacques, doomed, irredeemable from within? Or does the seemingly logical concatenation of painful, merciless, ferocious, unforgiving feelings which dominate this fabula numb a part of the soul leaving the other sunk permanently into a dreadful pit and making readers resign themselves to accepting with a shrug that "That is life"?

But if the naturalists' aim was to present the "whole of life," can we accept this one-sided picture in which subterranean forces overpower the individual as giving us the whole? Maybe the beauty of human life within, as it projects itself from the elevated spheres of the noble, creative spirit into culture does not, after all, encompass the totality of the factors which enter into the human predicament. However, it is obvious that primitive and brute forces — although their interference is undeniable — do not in their workings even approach the functional level from which the authentically human significance of life surges. The "human beast" — except perhaps in some rare pathological cases — does not represent humanness as such!

To maintain a "naturalist" stand one has to make a tacit metaphysical assumption, namely, that by "nature" should be understood only the organic, physiological functioning of man and whatever is mechanically dependent upon it — the lower circuits of the empirical soul. But such an assumption is unwarranted. These circuits left to themselves would not account even for the inventive and calculating reason that is indispensable for conducting a human life at any level. The reason which plans and invents belongs already to the functioning of the human creative orchestration, the fruit of the womb of nature, which nature prompts so that it may expand through the specifically human avenues of life; the proficiencies of the subliminal circuits of the human soul tame the surging of our brute drives and bring them into a network of specifically human, creative tendencies which promote full *human* existence.

This quick excursion down some avenues of different literary inspiration yields an instructive contrast, one which clarifies definitively our understanding of the passions. In our use of the terms "passions," "passional," "subliminal passions," etc., we attribute to them pitch, force, power, etc. and we describe them with qualifying terms like "vehement," "frantic," "raging," "indomitable," etc. Now these same adjectives are used for primitive urges as well as for the subliminal passions. Still, it becomes clear that the term "passion" is reserved exclusively for the latter. Although in general usage we indiscriminately confuse these different forces, we understand by "passion" *a sustained, all comprising, and purposefully oriented inner force that drives us onwards.* With this distinction made, it becomes obvious that only subliminal imperatives attain the sustained intensity and the purposeful, oriented, continuing dynamic force that is characteristic of passions. It is also clear that if priority is not given to the subliminal manifestations of the human being, we fall far short of the goal of giving the spectator, reader, hearer the "fullness of life."

And is not that the aim worthy of literature's vocation?

(e) *Recapitulation*

The reader requires literature that lifts him/her out of everyday attitudes. Literature is great when it has, more than an involving plot, an intentional structure. It is a pattern of intentional knots of significance that makes up the fabula ciphered in the script. When pursuing the sequential unrolling of this pattern in reading or viewing, and while contemplating it in our memory, we decipher the fabula, and whether we hit upon the essential point at which the core significance of the work emerges or only at a random point, of decisive moment for our retrieval of the significance of the work is its striking a chord within our experiential system at the deepest circuit of our subliminal soul. This entire system is then ignited and simultaneously responds and acts. The now vibrating stirrings of our subterranean "depths" are ready to be engaged by the call and urgency of the situation in which we find ourselves, be it in "real life" or in that situation into which the author has put his protagonist; in the latter case, the reader will retrieve him from it in a powerful and aim-oriented response to the provocation of his/her vital urgency, with all of his/her endowment being engaged in this response, with all of his/her imagination at work.

When pursuing the sequential unrolling of this pattern in reading or viewing, and while contemplating it in our memory, whether or not we hit upon the essential point at which the core significance of the work emerges or at a random point, of decisive moment for our retrieval of the significance of the work is its striking a chord within our experiential system at the deepest circuit of our subliminal soul. This entire system is then ignited and simultaneously responds and acts. The now vibrating stirrings of our subterranean "depths" are ready to be engaged by the call and urgency of the situation in which we find ourselves, be it in "real life" or in that situation into which the author has put his protagonist; in the latter case the reader will retrieve him from it in a powerful and aim-oriented response to the provocation of his/her vital urgency, with all his endowment being engaged in this response, with all his imagination at work. In fact, the reader/viewer simultaneously engages these stirrings in the form of a formidable *elemental* response ("*elemental*" *because the work imaginatively prompts out of his powers at their very limits and out of their elementary simplicity a most ingenious inventive response*); he responds to them by means of the elementary forces of his subterranean natural endowment now molded by an innermost *imaginative passional* reaction. This reaction means a *unique transformation of the elementary stirrings*, one which is wrought by the infinitely varying play between the situation, on the one side, and the inventive imaginative resources of the human being, on the other.

This movement in which the question and the response simultaneously surge up to confront each other leads to a uniquely crucial development within the human individual, one that reaches down to the very groundwork virtualities of the human condition. What is here at stake is the basic capacity for transformation within Nature itself, from the numb elementary stirrings to those informed by the imaginative interplay of the three basic sense-giving factors: the aesthetic, the moral, and the intelligible senses. At the level of transformation that is informed by imaginative sense-giving, our animal nature — the animal psyche, the natural soul — is partly illuminated by the intelligible factor, and thereby elementary formless and bruteness acquires the *specifically human* emotional coloring. It is lifted to a differentiated, milder, and softer tonality where sentiment for others, beyond the crude egotistic instinct for survival at any cost, prevails. The sphere of this higher tonality, which marks the advent of the human sphere in the individual, I have called the "subliminal" sphere, or the "subliminal soul"; its powerful prompting forces and tendencies I have called the "subliminal passions of the soul," in contrast to the brute natural promptings and violent forces that agitate us so blindly.

Now, since in the just described real life encounter between these vital forces and human imaginative responses to them, there emerge the imaginative elements, and since in the response of the human being the elemental passions are released raising the tonality of his experience above that of a strictly natural being, so the author who has in his work incarnated his vision and carefully prepared means to reach and stir the deposits of the subliminal soul operates at this level although, as pointed out above, he penetrates even deeper into the elementary realm. He operates, in fact, through the type of response that he aims at in order to effect a profound transformation in the reader.

Yet, this reaching to the very depth of the human being and thus being able to strike his deepest chords and play upon them and transmute their pitch and range and cadence is possible because whatever vision of life, of the world, of the human predicament wide or narrow that the literary work crystallizes, it is centered upon the human being, and through that focus on the human being it, so to speak, makes a person-to-person appeal, allows for our identification with its deepest motives, moods, hurts, sufferings, longings, with defeats and victories, with upsurges of independence and the necessity of having to submit to alien forces. Indeed, the innermost self of the human person, caught in doubts, uncertainties, the darkness about itself, and about the world around, and about its destiny, resonates with the heartbeat of the work. The heart of the reader beats to its rhythms.

It is the innermost life of the person involved in the conundrum of its life-struggle which holds him throughout the course of the novel so that it touches him most intimately. It remains to be asked: to what avail? Indeed, this first movement of expansion from within, captured and molded by the writer in special evocative generative fashions, brings forth from the subterranean zones of our soul the drowned and silenced voices of the flesh, of the heart, of the spirit's longings, and a yearning to vibrate with the life of the present moment is mirrored by another movement in which the imaginative charge of the author's vision is released, its topsy-turvy, blind, uncoordinated whirl acquiring perspectives on the wings of imagination, perspectives which find a place with the domain of the life-simile that the author prepares. That is, the author's vision brings in the sentient emotive links, prompts new fusions of emotions, and brings about the enhancement of some moments left in the shadow, thus bringing out soundless, shapeless, indiscernible, and numb elements which otherwise obstruct or energize, without our knowing it, the emotive/sentient arteries of experience. These are lifted onto a plane *illuminating* their obscure contours from within through exposure to the sentiments which now surge. In short, the writer's vision does not offer the "light" of intellectual understanding nor the evidence of "truth." Rather it expands the inner sphere of the recipients beingness.

(f) *An Excursus on the Priority of Enjoyment and the Futility of Using Any Single "Method Of Interpretation" Exclusively in Analyzing a Literary Work*

From the foregoing it is clear that the key to the life-significance of the literary work should not be sought along the lines of intellectual understanding but rather, along the lines of *poetic enjoyment*. Poetic enjoyment is, indeed, the factor of the Human Condition which — as we have expounded before (cf. "Enjoyment and the Poetic Sense," *Analecta Husserliana*, Vol. XVIII) is responsible for the fullness of psychic acts and brings to them the "poetic sense"; this latter lifts the tonality of experience above strict survival interests. Beyond that, the poetic sense accounts for the creative imaginative amplification of psychic acts which occurs in the transposition which endows the novel arrangement of events and situations with a powerful — as if magic — appeal: they become "enjoyable."

Artistic works are attractive simply because they are enjoyable. Musicians, poets, bards, painters are prompted to exercise their skills not only from within, that is, out of a personal drive to create a new reality, but by the public too, which is eager to enjoy their work. When we now read or watch enacted Greek tragedies or comedies, we use all acumen to "understand" them, chiefly because of a certain cultural distance. These same plays, however, were originally consumed by a general public for their direct appeal, the plays of Shakespeare being the well-known example.

This direct appeal to enjoyment focuses the reader's or spectator's attention and his retrieval of the work's meaning, though some cultural prerequisites are certainly assumed and a certain, minimal degree of education will enhance the enjoyment and make the marveling and wonderment more meaningful. There is, of course, to be considered the many-layered structure of the artistic or literary work through which we have to make our way in order to advance in our discovery of its meaningfulness. But although studying the intricacies of structure, the specific devices of the respective medium, the various modalities of a work's crystallization, the means used to establish the script, the techniques applied, etc. certainly gives us extensive "knowledge of the script and of its crystallization, all this only makes a contribution to our enjoyment; it is not what makes it surge.

Enjoyment is precisely our spontaneous response to the fable in a script, on the canvas, in a sequence of movements or events which surges from out of the entire maze of the human experiential schema in an unpremeditated way. Only in this spontaneous way in which the reader or the spectator naturally summons all his wits and sensibilities in answer to the script's call for unfolding does he find a path through the entire conundrum of the work's structural entanglements, the intricacies of its linguistic or artistic techniques, etc. while focusing on the guidelines which imaginatio creatrix conjured within the work itself. Again and again the reader conjectures the intentions of the author, and on how the author's subjective situation might have left a stamp on the fabulation and on how knowledge of this might contribute to the appreciation of the work and to the clarification of many points concerning its message. Nevertheless, the spontaneous effort made in the work's reception, the work of this spontaneous response cannot be replaced by any studied reading.

Life-significance, we have attempted to show, does not proceed from and does not aim at the intellectual faculty of the human being. It expresses the innermost passional concern with, as well as allegiance to, life experiences. The vision of a great work has to be retrieved in the sweep of the subliminal passions of the soul; no elaborate study is capable of a successful launch into this space. Only aesthetic enjoyment gets us off the ground.

Hence, it is clear that we cannot single out one "method" in order to "apply" it in the appreciation of a poem, novel, play, etc. without thereby introducing a misleading and distorting bias into our appreciative experience. In any intellectual analysis of a script, taking care to avoid such distortions would mean consideration of all the lines of investigation possible — which would amount to an inquiry so extensive that the spontaneous bursting forth of the work's vision would be stillborn. Let us, in contrast, follow the guiding points which the work itself offers and follow with all our passional response the aesthetic enjoyment of the work. In doing so, we cannot miss its peak.

To summarize: the purpose of the literary work is, above all else, transmit the vision which the author has incarnated within it to the recipient-reader. First, the work establishes contact with the reader within the life-simile of its fabulation; then its deepest intuition can resonate with the reader's own experiential system making it pulsate now in his very flesh and blood. Thus, the vision which the work incarnates may reverberate within the recipient-reader's personal life experience. Then, in a further step — which is, however, simultaneous with the preceding ones — the vision of the work expands the experiential schema of the reader opening new vistas that force him out of his self-enclosed cocoon, new perspectives for his feelings, thoughts, and imagination. Lastly, and most significantly, it illuminates his innermost sentiments, and his appraisals of and approaches to life.

It is in this last all-embracing step that resides the life-significance of literature. It is only in enriching our spectrum of life appreciations but mainly in raising them to a higher tonality and differentiating as well as expanding our range of experiences, swinging from one possible extreme to the other, that literature fulfills its vocation.

SECTION II
THE PASSIONS AND THE ELEMENTS IN THE HUMAN CONDITION: LIGHT AND THE SEA

CHAPTER ONE

THE PRIMEVAL LIGHT AND THE BIRTH OF THE LIFE-WORLD

(a) *Light and the Essence of Living Nature*

I have anticipated this inquiry in my prodigal use of the metaphor of light in its various modes while discussing the twilight zone of the soul. Our language in discussing almost any subject is suffused with metaphors of light, and we can say that empirical life is conducted with constant poetical references to light. We will limit ourselves here to merely sketching this usage's metaphysical nature.

From the point of view of the metaphysics of life, the fact that the book of Genesis in describing the stages of the origin of the universe gives us at the strategic moment the Creator's command. "Let there be light" discloses a crucial philosophical discovery. The emergence of the primeval light within the schema of the creative generation of the world simultaneously brings out the most fundamental principle of the cosmic as well as the transcendental constitution of the life-world and the first emergence of the vision of life itself, life as the play of Nature or the quintessence of Nature. It is no wonder that man's creative imagination radiates in all directions the poetic life-significance of light.

In fact, the meaning of light's primeval or, better, primogenital status can be understood in a literal sense as its being "first in the line of generation." It can be seen that the element of light as such brings together the three seemingly separate (and in their far-flung distantiation difficult to bridge) spheres of Cosmos, Nature, and the human inner self. Light is, first, the dispenser of warmth, and with its emergence Nature receives one of the major factors promoting generation, even as an increase or decrease in the dispensation of warmth puts life in question. (And yet generation and then growth and decay mean life itself, the principle of Nature). Furthermore, light in all its degrees,

A-T. Tymieniecka (ed.), Analecta Husserliana, Vol. XXVIII, 77–85.

from the zenith of brightness through all its lesser shades down to darkness synonymous with its total absence, brings about the pluridimensional expansion of what we consider the world around us and the world within ourselves. Secondly, the expansion of space without illumination — the mere extension of matter — if possible at all, would not allow either the unfolding of forms of life other than the most primitive or the life-world of man with its full variety of other living beings; man's active expansiveness within this vast range of natural perspectives as well as within his own constructive creative extension of the avenues and forms of living would not be possible.

With the emergence of light the vastly varied pluridimensionality of the world as a locus of life unfolds. The world of life which is the earth's surface is the midpoint between the levels of the atmosphere above and the dark strata or ocean depths below. And in the midst of the indescribably colorful, varied, changeable world of ever advancing shapes, volumes, and movements is the turmoil of life with its ideas, trends in tastes, appetites, expressions of feeling, and desires. It is the primogenital light which brings us glare and shadow and which divides the moods and the inward psychological dimensions of experience into present and past, and which, first of all, sets horizons for our mobility as well as our repose.

With this we have moved from the dimensionality of the world of life to its vision.

(b) *The Primeval Light and The Emerging Vision of the World of Life*

By what we have said about light we have already set forth some of the co-generic points which make light in its power to found the world of life and the specifically human world convertible into its essential partner or effect: vision. Indeed, when we consider the emergence of the dimensionality of space, the distantiation of spaces, the differentiation of shapes and of qualities like color, fragrance and texture, and, even more, consider the differentiation of things and beings in Nature, we already touch upon the essential factor of experiencing proper for living creatures: vision. There is a good reason why among the ways of acknowledging these qualities or the means of constituting them for recognition by living beings, ways that belong to several types of basic sensory experiences, e.g., touch, taste, smell, or hearing, science and philosophy have attributed primacy to sight.

We may undoubtedly encompass a pluridimensional mass by the

sensory experience of touch, but we may then only identify or circumscribe an object insofar as it is accessible to the extended hand. We cannot reach much with the sense of touch; likewise we experience only within a limited circumference with our sense of smell. It would seem that since hearing can occur well beyond the source emitting a sound and that since voice extends our reach into space far beyond us, then sight has a worthy competitor in hearing. However, the resounding voice does not open a horizon; neither does it expand into infinite space; on the contrary, the final form of the resounding voice is the echo. In contrast, sight extends to all sides and is not limited by the contours of the specific perceptual object of this singular perception; it extends also above each individual perception to take in the horizon of a world which seems to be the circumference of perception as such. This horizon itself and this expansion in space on all sides constitutes a *vision.* It embraces a visual field with the self as its center, a center which moves as a person moves so that the visual field remains ever around him.

Not only does sight as a sensory experience of the human being as well as of other higher living beings spread the living individual into innumerable, if not infinite, perspectives beyond himself — perspectives within which he may differentiate other beings as well as the entire spread of living nature — but it allows him to "situate" his own position among them. Furthermore, and what is of crucial significance for his life-world existence, he may due to sight 1) centralize his position in any field no matter how uncharted without getting lost in anonymity to himself, and 2) accomplish this naturally because the structure of sight experience is such that it synchronizes visual perspectives in such a way that he can gather them all "around" himself "as if at one glance." It is obviously due to these two proficiencies that sight plays a primordial role in our cognition/constitution of the life-world — not only for humans specifically but also for animals bestowed with it. Although our world and life experience as such is obtained through the cooperation of all of our sensory reactions, yet the overall task of providing a "picture" of the living world in all its dimensions is accomplished mainly through sight: sight allows us living beings, especially the human being, a global *vision* of the world of life. To this latter belongs also a principle which is of the greatest significance for man's situating himself within the life-world and which belongs also to the nature of the world of life as such, namely, the horizon. The human being envisages the place within the actual world in which he always finds himself within the confines of a horizon. The animal always establishes a "comparable horizon" for his circumambiant sector of the life-world. But it does this through smell rather than sight, and the expansion principle of the horizon belongs to vision.

With these considerations, we have entered directly into what the sense-experience of sight is really about: vision. My "vision" of the world does not cease at any horizon that my movement projects; on the contrary, beyond any horizon there is implied within my vision an expansion into infinite space, or space which is there. We might have learned from science that our earth is round, and we may rationally conclude from that that the spaces upon which our actual horizon of the world borders are physically calculable, measurable, etc. And yet, there belongs to the experiential significance of vision an experience of the world as extending into infinite space, as commanding a feeling of repose and openness, infinity.

To finish this brief discussion of the particularly "physical" aspect of light, let us emphasize at this point, first, that from what we have already said it is clear that it cannot be subsumed under the sensory performance of man which we call "seeing." On the contrary, seeing appears to be an indispensable outcome of the role that light plays in the origin of the life-world, its differentiation, and the establishment of the life-system and its flourishing. This role makes it "primeval" and gives it a status such that the emergence of the organ of sight in living beings assumes a secondary position in the line of the genesis of life and, furthermore, a position of adaptation to life's primogenital demands, the position of an organ entrusted with a specific role in life's progress.

Second, we have to specify that it is not only by means of the experience of seeing, which in itself is merely an intentionally restricted and directional experience, that we attain this access to the ascertainment of the primeval light as such. Even this experience in its abovementioned fullness benefits already from the specific orchestration of the living agency. It seems that man shares this experience with the majority of living beings. That we humans may reach beyond our experience toward its encompassing conditions, filling out the steps which lead to it and which are not directly experienced is certainly due to the creative orchestration of our faculties with the Imaginatio Creatrix at its center. To reach the notion of the primeval light in the book of Genesis, to either receive it revealed to or to invent it as a mythical response to his inquisitiveness man had to have worked it out within the subliminal workshop within which all of his faculties emerge in full, drawing creatively upon his entire endowment. This means that we had long already been dealing with the ELEMENTAL LIGHT.

With this we have prepared the groundwork for the full-fledged appreciation of light, for what we may provisionally characterize as "inward/outward vision."

(c) *The Elemental Vision*

Strangely enough, being oriented "visually" outward, out of ourselves toward the exterior, we rarely realize that our individual and personal life which goes on incessantly within us throughout the life experience — this pulp of our existence through which we experience ourselves to be full, and not void, as having weight and not floating, as a substantialized embodied being, that we identify constantly, at every instant of our conscious life, constitutes within us global inward vision. We do not, in fact, lead this inner life of ours "in the dark." It is not only that the actual field of consciousness within which the experiences of every instant appear "in the full light" of consciousness is always within a sphere of inward illumination, but that the experiences of the present moment on receding into the past enter into the sphere of the "shadow" until they vanish from sight "into the dark," and that our tentacles into future experiences reach out into vague but illuminated spheres. But what is of the greatest significance is that these zones of luminosity present only fragmentary circuits of radiation from the overall inwardly radiating vision that somehow gathers to itself our living beingness *in toto*, the entire circumference of life experience. We will come to this totalization of our beingness later on.

Moreover, do we not consider the sphere of the life experience of which we are clearly conscious to be a sphere of luminosity? Indeed, there is certainly a sort of interchangeability between the clarity of consciousness that reigns within our inward self and the "clarity" with which we are present to ourselves and envisage our life enactment, conduct, motivations, hopes, and projects. We may therefore be tempted to, as is usually done, identify a degree of self-consciousness with lucidity. A confused consciousness is considered to be "unclear, nebulous." From this it is only one step to considering a "clear" consciousness — in its clearest form presented by the works of human intellect — as *the light of the human mind.*

Thus, the inward vision which is basic to the unity of our living beingness — the work of the subliminal forge in which our organic/vital, gregarious/psychological, and specifically human factors of significance come together within a new mold is, in truth, not self-enclosed in any way. On the contrary, it is "*inward*" *in that it is capable of prompting our beingness "out of itself" — ourselves — through the sensory perception of sight* as well as through an entire system of intentionality, and above these through our subliminal longing and quests. Thus, this outward vision of man as situated within a world expanding in infinite horizons is simultaneously the fruit of these subliminal workings, and it presents, simply, the expansion of the inward vision, which in turn draws upon the outward vision. In short, human elemental vision, the fruit of our unique subliminal workings, which draw upon our entire organic, vital, psychic/gregarious, and creative virtualities, *unites them all within a luminous emergence from brute nature into the human world of life with its infinite horizons.*

Indeed, it cannot be stressed strongly enough how amazingly we live within a sphere of luminosity, or within an orbit of light into which the luminosity of our inward, psychic realm naturally extends. Our inner sight reaches out into the luminosity of the "outward" life-world of our psychic vision. Human life is as such all-luminous.

(By living is meant here our awakened state in which, as Heraclitus said, we lead a communal existence with others; for when sleeping — on closing our eyes — we distance our selves from active communal existence, we retreat from it for a while (and also from the luminous zone of our psychic life as a field of active life-promoting involvement) into a passive sphere of our beingness, away from the world, a dark zone where in our dreams a luminosity is experienced that is not the light of primary, active, alert life functioning but, rather, a secondary, memory-induced projection.)

(d) *Intelligence — The Light of the Mind within the Human Creative Orchestration*

As we have pointed out before, it is the faculty of intellect which within the luminous sphere of life experience spread by consciousness around the pole of self-consciousness as it gathers all our experiences, their performance, their motivations, and their forces together into the unity of our beingness, establishes, through networks of rational connectedness and structural linkages, the cohesion of purposeful interlinkage and cooperative fusions. This interlinkage which was operative in all the circuits of organic/vital life progress is, with the advent of the creative orchestration within the Human Condition and the surging of the four faculties which undertake the construction of a specifically human world of life, endowed with man's unique significance. In fact, imaginatio creatrix, memory, will, and intellect could not function in "darkness" without the "light" proceeding from the workings of the intellect as it is incorporated in the molds of consciousness. It seems that in order to advance from the opaque organic/vital existence in which the rational interlockings are operative within the constructive progress of life at varying levels of complexity — at first, only under the entelechic direction of the principle of relevance and not through deliberate selection and choice — it is necessary that this complexity unfold to the point of installing its own intrinsic central agency, an agency that would centralize the automatic processes of the work performed by the separate organs by incorporating them, spreading out through them while at the same time distilling from their innumerable functional circuits and their interlockings a specific synchronizing circuit that functions as if disembodied from the singular networks of operations that it runs through, as if gathering them all in their "pure rationality."

The individual life agency of animality already manifests a type of agency that synchronizes the networks of functional circuits and their corresponding relevancies. But only with the advent of the complex constructive phase of the Human Condition do we have a synchronizing complex with so tight a hold on the reins that really controls the entire range of the organic, vegetal, psychic/gregarious processes; moreover, it is with the crystallization of its virtualities at its advent that the Human Condition brings about within the creative orchestration the emergence of the intellectual faculty.

What does the emergence of the intellect mean besides that within the surging of the self-awareness of the individual within the creative orchestration that embodies this synchronizing effect there is performed the novel twist of flexibility, variation of agency, choice, freedom of decision, etc. — the dawning of awareness of this possibility within all our acts as well as in all our operations? With the taking possession of these operations in a glance, a "light" is introduced into the otherwise opaque, nebulous circuits that otherwise have at best a slight measure of flexibility. The light that self-awareness brings is not only the inward-outward reaching subliminal vision of the individual life, but is the vision of the self's "roving eye" which moves at one with the network of the individual's operations — the network of their rational connections clad in innumerable experiential molds —which becomes transparent to the eye, operations which he can oversee, scrutinize, and control. We live within the luminous zone of our consciousness which gathers together our entire beingness. The intelligibility of the rational inter-lockings in the phenomena of our inward as well as outward projections is now exposed to the light of our intelligence.

(e) *Light, Sign, Word*

The dawning of the light of intelligence means in fact the bringing of various degrees of luminosity to our beingness, luminosity which culminates in the transparency of intellectual consciousness. It is this role of the light of consciousness that accounts for the mind's piercing all the otherwise blind and numb strata of our existence within the world and gathering to itself the differentiated shapes, dimensions, colors, and textures of the things that bear them in a way that allows their differentiation as elements within a unified field — that of the life-world. Now, this differentiation can be grasped by the mind "in a single glance"; one is no longer lost in the opaqueness of all these elements and their articulations. Our mind masters this differentiation, and we utter specific sounds and words to hold them before our mind: names that betoken categories and classifications. *Thus, light is the subliminal origin of human reason and the word.*

We have indicated how the "light of the mind" illuminating the inward zone of our psychic functioning is critical for the sharp synchronization of functions which allows the accomplishment of the most complex operations. Its culmination, however, namely, the expression of this synchronization in the voicing of a "sign symbol," in the incarnation of the meaningful construct of a word cannot be stressed enough. The "sign/word" or the "significant word" is upon our analysis the final fruit of this infinitely complex synchronization of relevant functional links that is the emergence of the mind as a faculty synthesizing them all "at one stroke." The word bursts forth from the infinitely complex linkages of life's constructive processes as the decisive factor in the luminous intelligibility of life. It might be life's definitive and, in this sense, ultimate manifestation, for it announces life's complete genetic accomplishment. Thus, in passing from light to the word, we cover the infinitesimal distance spanned when the cycle of life's genesis was completed.

The word as the synthesizing outcome of all the functional circuits of life that carry the individualizing progress of the human being means their breaking through to light, a decisive step in this progress. The word is by no means the automatic effect of this progress, as we can see in the case of the human being on-the-brink-of-existence, who finds himself in this situation because his ties to communal life have been loosened and he is despoiled of this network — the intelligible links of collective luminosity which hold human beings together. It then costs him an effort to utter the word, and the proper word. The word, appears, in fact, to be the key to his re-establishing these existential links, that is, his sanity. Incapable of uttering it — of participating in the luminous zone of existence — he sinks slowly into the dark circuits of the psyche and deteriorates.

CHAPTER TWO

THE PASSIONS OF LIGHT

(a) *The Subliminal Passions That Tend Toward Light: To Understand, Encompass, to Create; The Passions for The Ultimate Quest and for Infinity*

As plants direct their "gaze" toward the sun, so the human being tends toward light. We may, in fact, discern a host of subliminal passions which spring forth from the subliminal workshop in which the light of the mind crystallizes and releases its powerful dynamic force. Oriented toward the light of the mind, they tend to illuminate, to penetrate the obscurity of life, and to encompass in lucidity life's circumference, to radiate; in short, they glow with the rays of the Logos of Life. This spread of subliminal forces bursts out from the specifically human womb where the human being is made human and culminates in the passion to create, to bring to light, to manifest an all-embracing vision of life in the form of exemplary works of art, of thought, of reflection and technique.

We will close this investigation with a brief discussion of the creative vision as the culminating term of the passions of light, brief, because we have devoted numerous studies to this topic.

Yet we will be led into it by an analysis bringing forth the passions of light. Indeed, the entire constructive progress of the human being seems to express a drive toward light. Since the clarifying of his procedures, his selection, his motivations, his capacities only comes to man when he attains the self-awareness of the "eye of the mind" and by the same stroke enters into the possession of his powers, which then increase a hundredfold, so it is the maintenance and further unfolding of these propensities that is his constant task. Indeed, specifically human self-individualizing progress — man's self-interpretation-in-existence — is conducted already within the zone in which the subliminal passions spring into action. To devise new avenues for his existence which calls for a constant effort to *understand* other human beings within their own luminous private zones, the human being calls upon his driving passion to *understand* and to know, to *penetrate and to encompass*. This keeps

A-T. Tymieniecka (ed.), Analecta Husserliana, Vol. XXVIII, 87–99.

the human being in existence; the passion to know and to understand does not stop at any particular objective, neither does it stop at the matters that concern the business of concrete individual life; man's luminous zone of existence is kept alive and man is kept abreast of actual happenings by his fulgurating passion to know and to understand. He not only looks, hears, tastes, talks to others, writes to them, etc. with the innermost intention, ingrained in everyone in his life-enactment, to "know," to be informed, etc., but he gazes far beyond the circumference of the practical life. He does this naturally since the practical life is itself immersed in the ever-expanding circuit of civilization — on the wings of the expansion of man's passion to know and understand — which does not stop at the practical steps which it takes. On the contrary, the passion to know and to understand further and further spheres of what surrounds us, of what carries us, of all that concerns us tends to infinity and is what prompts the scholar, scientist, builder, and inventor to agitate his faculties so that the searching eye of his mind may gather in all about it and he may push into further and further territories attempting to throw light on their obscurity and discover the lineage that leads to primeval Logos. We may say that the Logos of Life along whose rays the constructive progress of life proceeds becomes itself the objective of our passion to know and to understand; it is to the Logos of Life that we tend in our never ending quest to illuminate. Doing so, we make the Logos of Life radiate through our practice within our luminous zone of existence. We make its luminosity expand.

It appears now that the element of light in its expansion from the subliminal workshop into the mind's intelligible working — through the fulgurating images of the creative imagination, the illumination of emotions, feelings, moods, etc. by intelligible "sentiments," the emergence from experience of a "view," and "insight," a "cognition," a "constitutive" coalescence, a decision, the realization of the intelligibility of moments and of the relevance of moments to each other — is the essential feature of experience by means of which this wondrous translucent sphere of human existence emerges and maintains itself. Its seemingly most direct constructive progress appears to occur through the works of the "pure" intellect: through our reflection, thought, deliberation, judgement, calculation, etc. However, the element of light is not caught in the workings of the intellect as mountain water is caught in a stream; on the contrary, it is only by its avoiding being usurped by pure rationality that it spreads into all aspects of human functioning, to wherever human experiential tentacles reach. It cannot be caught by "pure rationality" alone. Far from it.

It seems as if the constructive progress of life tends toward light, more and more light, as if ultimately human life were suspended upon light. Indeed, in this respect, we have to observe a point that is all too well known and has been discussed above yet merits being pointed out again here. It appears as if the subliminal light along with its radical opposite/contrast, darkness, serve to demarcate life's zones, light marking the zones of growth, health, sanity, and darkness, the zones where the virtualities that ascend to the level of life brew and coalesce to either surge up and emerge fully into the light or to fall back and disintegrate. The ancient mythologies saw in the opposites of light and darkness metaphors for health and annihilation, life and death. The opposition between light and darkness in its subliminal primogeneity also comes through in mythology in expressions of the opposition between the forces of good and evil — Ormazd vs. Ahriman, heaven vs. hell, angel vs. demon, etc.

At this intersection of life's basic interests, health and good, a most complex host of metaphorical conceptions and images, found in various forms in all cultures, is situated; creeds, the arts, literature, and the concrete experience of mankind feed upon this feast.

We will in the third part of this study probe further the zone of darkness that lurks at the threshold of everyday life as it is explored in recent literature. At this point, we have still to bring out the culminating moment in the human quest for light.

(b) *The Passion to Understand — "to Bring to Light" and the Role of the Word*

As emphasized above, elemental light appears to be an essential factor in experience *per se* as a conductor of its intelligibility. And at this point, a most important insight surges. We have in our previous investigations attributed the finding of intelligibility to the operation of the intelligible sense as a primogenital factor of the Human Condition and as promoted by the human faculty of the intellect. At the same time we have attributed to the intellect the distribution, creative expansion in form and function, promotion, and manipulation of intelligibility. In the new perspective being developed here of the primeval as well as subliminal genesis of light, *light has a most essential convertibility with intelligibility, a convertibility which it displays in its life functions; thus, we come to a point at which intelligibility has to be recognized as, first of all, not being the product of the faculty of the intellect.* That is, although the intellect promotes it and handles its unfolding and extension, it does not generate it; it only takes it over from the virtualities of the Human Condition while entering into their operation.

Secondly, we have to distinguish *strains of intelligibility other than those which enter into the workings of the faculty of the intellect.*

As a matter of fact, it appears as if there is a vast spectrum of strains of intelligibility infused into life's innumerable ways of functioning beginning with its very first individualizing steps and then advancing through the entire scale of the evaluation of relevances through which the constructive progress of life proceeds. It is, however, within the creative orchestration that emerges with the advent of the Human Condition that intelligibility, in all the degrees of lucid transparency which the Logos of Life establishes in its rational constructive articulations, arrives at its culminating point. It surges through the entire spectrum of human experience in all its modalities within the mind's zone of luminosity, which luminosity it brings about. It is undoubtedly the translucent and tightly articulated intelligibility of what has been traditionally called "pure reason" established and dominated by the working of the intellect, the architect of life's progress. And in this form it undoubtedly forms the skeleton of life's expansion, of the translucent sphere of human life, besides offering a regulative mechanism and pattern by which the human agent may conduct his existence.

However, and here we come to an important point of our discourse, this skeleton does not control the entire spread of intelligibility within human existence. The radiating intelligibility of human experience flows over this rational skeleton on all sides, steering the vehicle of life. Furthermore, the passion to know and to understand is directed as much toward our inner self as toward the surrounding world. In fact, it is this passion to know and to understand our actions, feelings, relationships, motivations, etc. that is the intrinsic force of our inner development, our character building, our self-interpretative life-course, as well as of our pursuit of our transnatural destiny, and lastly of our turning against life-interests and turning toward the religious quest.

This explains not only why we are always for ourselves the most interesting subject of conversation, but also why we seek to clarify ourselves to ourselves by bringing to the light reflective images of ourselves in autobiographies, memoirs, diaries, confessions, family histories, genealogies, etc. and why we seek to plumb our deepest subliminal empirical psyche in psychoanalysis, which once embarked on will never lose its fascination, and why we always talk about our past as well as that of our families and friends; we are bringing all these things up from the dark spheres of our psyche into the luminosity of the significant word.

In fact, since nothing within the impervious strivings of individualizing life remains isolated — on the contrary, all prompting forces, functions, tendencies operative within the framework of life's unfolding remain in a cooperative network and only as such enter into operation — but every pulsation of life mingles with others in the common task to advance life's constructive progress, so this urge to pierce through all the constructive circuits and acquire a form of one's own in the full luminosity of human individual, as well as collective, consciousness, and to bring "into the open" our personal experiences as well as collective deeds is of paramount significance for life. In fact sharing events of our life, giving them formulation in words and concepts, "frees" our innermost psychic states, the personal feelings, doubts, and reflections that are the results of accumulated raw experience and which have great power to overwhelm us with anxiety, anger, fear, anguish, regret, remorse, shame, sadness, etc. which being buried in the dark zones of our functioning lie there as a heavy burden and a negative influence upon our progress, a situation that calls for light and unencumbered flexibility of mind. Sharing them with others requires an interpretive effort, a disentangling of their instrinsic concatenations, and a hunt for causes and hidden motivations, and that means to bring them to the light of our clear consciousness. Sharing them also means transposing them from the limited frame of our own luminosity to a collective realm of understanding and knowledge. By doing so, we rid blindly thrown together raw complexes of their accumulated power; we exorcise them, so to speak, of their hidden claims to menace our carefree advance; bringing them to an open field within a large collective network, we are able to release acts of reconciliation within ourselves, forgive ourselves our past errors, willful wrongdoing, etc.; we are able to appreciate too our own and others' congenital weaknesses, indolence, incapacity to handle this or that situation, and, lastly, to recognize the limitations and weaknesses of the human being as such.

No wonder that, since the word ultimately synthesizes the entire spread of the functioning which constitutes the human being and is the key to man's existence (individual and cultural and societal as well), it is the tendency of contemporary psychoanalysis and philosophy to get at the root of the human by taking apart all the components of the specifically human life, by making the word the target of scrutiny, for the word contains within its meaningful content the sedimentations of all of cultural, societal, personal, and individual existence.

A similar "healing" role of the word, a "bringing to light" has naturally been recognized and practiced throughout the ages, and been a constant in the collective consciousness of societal groups, nations and the whole of humanity. Bringing "into the open" criminal acts and other wrongdoing in a society and discussing "in the public light" their occurrence, reasons, effects, and guilt and innocence, and penalties airs the circuits of hidden ominous energies which if left in the dark corrode a society. There is no way to illuminate and clarify the dark zone of personal or of collective guilt in such horrors as war crimes other than to bring them into the public eye and submit them to scrutiny so that the eye of the mind may penetrate the dark zones of guilt, resentment, vindicativeness. It has an effect of healing the wounds inflicted by the crimes; in appeasing rage, it effects, if not reconciliation, at least a truce instead of implacable vindicativeness. This is the reason why nations dwell so long on the wars they have fought, whether won or lost. Innumerable are the accounts given years and decades after they took place. This is the reason to dwell upon and bemoan mass exterminations and persecutions long after even their direct survivors are gone. In fact, we deplore the tragedies willfully inflicted by man upon man much more than those caused by natural cataclysms. The healing which humanity draws from the light of the mind is one of the essential factors of our cultural life.

Accounts we have of spiritual progress and mystical states are also not, in fact, written as "testimonies" and read to give and obtain guidance or knowledge of how certain things are to be understood; they are, in the first place, the result of our passion to understand and thus bring to light in intelligible — enlightened — form the developments which take place within our obscure inner zones which are otherwise isolated from the light of others; it is this "enlightenment" which is the main task aimed at in the effort to penetrate and discover the intelligible elements in our inward development.

By the effort to bring out into the light the intelligible element of our life's innermost struggle, doubts, striving, failures and victories, patient pursuits, and deepest convictions in rendering in proper intelligible form, we ourselves not only radiate the light of the Logos into the human world, but become more transparent to ourselves. We also "grow in light." No wonder that the word has been attributed magical powers in human history or that it has played a crucial role in esoteric creeds, kabalistic speculations, and religion.

In fact man's innermost urge toward light — to cognize, to understand, to encompass and to reach the ultimate reasons of his very existence carries him toward a light that is beyond the intelligible rationality of the life-world's horizons: it is in the guise of an "unearthly" light that man projects his vision of the "trans-natural." The so-called "appearances" of the trans-natural surge within the dialectic of light/darkness, and to conjure, indicate, and celebrate the presence of the divine among the faithful, light is used in various ways (lamps, candles, torches, etc.) that symbolize the sacred.

It seems as if the entire spread of the quest of the human being to transcend the confines of his self-encircling natural luminosity — which ultimately does not yield the clarifications of its own reasons and which under the scrutiny of man's "spiritual eye" appears strangely opaque — is but a striving toward the Ultimate Light; it is the Ultimate, the unearthly Divine, the God that would bring light into the world.

With this last quest for the Ultimate Light, the human being turns his gaze toward a transearthly infinity without horizons.

Thus, we arrive at the echo of Genesis in John's Gospel: "In the beginning was the Word." Primeval light, primeval logos seem to be the initial factors of life in all its phases of constructive progress; and also to give direction to the yearning of the human being for the transworldly. We have attempted to sketch their intrinsic coincidence and their nuclear interchangeability in the immense process of the self-individualizing advance of life. Human culture has universally expressed this coincidence in the metaphors which seem to reciprocally identify one with the other. As a matter of fact, we have been using these metaphors in our presentation all along.

Before we close this brief study of the element of light in its crucial life significance, let us recapitulate by emphasizing some points on the the role it performs in life's advance that we must bring out.

Not only does most of Nature tend towards light, but light as clarity, intelligibility — the light of the mind — characterizes the specifically human phase of life. It is the attainment of the luminosity of self-awareness and of the knowledge of one's place, situation, forces, limitations, etc. within the world of life that the evolutive progress of life seems to have tended toward. But, this accomplished, it appears clearly that cognition, understanding, and intelligible expression are specifically human strivings. The urge to understand and to express is, in fact, one of the greatest motors of man's own creative passion/drive: the passion that allows him to expand the virtualities of the Human Condition and fulfill his humanness.

Indeed, the passions of light, while pursuing their own trajectories, partake in the impetus of the primogenital creative orchestration of human faculties, and they form various particular routes of enactment as rays of this vast field of "enlightenment" that makes civilization and culture. But their crucial accomplishment and role in life's advance is to be found in the creative vision of man.

(c) *A Passion to Transcend the Horizons of Life on the Wings of Invention; Man's Creative Vision*

The eye of the mind makes the human agent in his central self radiate in all directions with luminous transparency in such a way that this self at the center dwells in a universe of hand, eye, and imagination within his own compass, a sphere to him translucent and yet encircled by horizons, which, although movable and able to expand with his reach, are grounded in the unchangeable stable data of his natural and Human Condition. The irremediable succession of life stages following the cycle from origin to extinction, the confines of bodily action, sensory affectivity, and mental powers are there. Even the seemingly unrestricted flings of imagination have their bounds. In short, this wondrous universe of man, this marvelously translucent globe of his own making not only allows him to expand but also delimits what he can attain. For from within the wondrous circle which we trace, we find in the exercise of our constructive powers limitations "on all sides" coming out of our innermost make-up, for the very assumptions of the Human Condition, although they empower us to blossom forth a luminous universe of our own, still set very fragile foundations for it. Despite the extraordinary ingeniousness which we bring to our inner workings to bring that universe about, and sustain it and make it progress with the progress of life, taking all our powers, forces, and genius into account, this wondrous inner sphere of ours floats on the stormy surface of life and is no more stable and no stronger than a soap bubble. In short, when the eye of the mind scrutinizes and covers all the life horizons, within and without, the human being appears to himself to be limited and restricted within his own confines. He works with his beingness as it is, but he has not set its rules and its limitations; he sees it all now, and his striving toward light prompts within him subliminal yearnings to *understand his ultimate conditions and their reasons*, to understand what his existence within and above its limits is about; to penetrate into the deepest circuits of human existence *per se*. Awakening from the contentment, complacency, and intimacy of his stabilizing ingrownness in the networks of his life, he feels discontent with its confining routine, anger with the chains which he feels lock in his forceful desires; he wants to break them, to be free, to undertake on his own a search for the hidden mechanics of life's unfurling. Lastly, he conceives an unrelenting urge to surmount these limits which his own making imposes on him. The first manifestation of the urge to transcend the human orbit prompts the human being to create novel forms of the significance of life and novel avenues for human existence to tread. The urge to surpass our present confines coincides with the dissection of the significance of our world of life and with the aspiration to give it a new significance.

Even if in the final account this would mean no more than pushing the horizons of life a little further out, the human creator still, like a demiurge, experiences the exercise of a power beyond his limited exercise of life and takes up his own exercise of life in its fullness.

The urge to create — finds its expression as I have analyzed it in detail in several of my writings — in the creative vision within which man explodes into innumerable possible "universes," without limit, into infinite spaces with movable contours, into spheres so high up that his imagination even cannot fly there. Indeed, when the urge to transcend ignites the creative proficiencies of the human being and he soars on the wings of Imaginatio Creatrix, there surges from within him an inward vision which catapults him out of himself into the bailiwicks "above" his horizons, as well as into abysmal subterranean realms "below" all light — into infinite evolutive provinces of life, infinite on all sides; and yet the rays of this vision meet within his translucent self, and it sets him free. I have said so much about "creative freedom" elsewhere. Now, is the time to give to the discussion of creative freedom its final touch. Once galvanized, the eye of the mind knows no end to its searching, and man's hand follows the eye. It pierces matter, splits the atom and its elements, seeks the germinal forces; it brings light into the darkest corners of life's composition and illuminates its infinitesimal threads; it pursues life's progress through immeasurable stretches of time and measures the imperceptible waves of the evolution of forms.

How sclerosed, skimpy, meager, etc. are the imaginings of the classical times in which our humanity first framed an attempt to come to grips with his overall vision in what is called ontology, metaphysics, ethics, etc. Man had first to understand his own concrete domain, his very way of existing within the self-projected globe; he had to find its structures, its forms, its principles, its ways of behaving, its limitations; he had also to situate his efforts to transcend within his orbit. Hence, the speculation of antiquity, and the philosophy of the Middle Ages. The Renaissance's efforts to probe deeper into man's possibilities, however, shifted the focus slightly. But with the explosion of modern philosophy, the past speculations of philosophy, their whole emphasis, lost their vitality, their applicability to the new issues facing man.

Man had to focus first on the "objective" structures, rules, and principles of what appeared to him to make up his solid "external" world of life — which naturally bedazzled him with its earthly horizons and seemingly infinite skies, the cycle of the seasons punctuating the flux of life and the natural meteorological forces, the cosmic powers of the sun, the moon, the stars and comets, and all the forms of living beings. To seek understanding of them was the first task of the mind. Hence, the first great creative visions of men — the poets, playwrights, philosophers/scientists — were fruits of the purely intellectual speculation of minds that marveled at our life-world. Hence, the abstract intellectual purity of their first principles which floated in ideal spheres totally remote from the actual course of things within the life-world. However,with modern philosophy, which followed the Renaissance in its move toward investigating the living and acting man exerting his forces and powers in the life-world, the focus shifted from speculative marveling to improving further life's conditions. The eye of the mind was followed by man's hand and was moved by his desire to not only understand but also to make, to change, to improve, to build. This called for the awakening of all faculties of the human mind which had been sunk in a slumber. The creative vision no longer took inspiration from the marvels of the world alone, but was ignited anew by a pressing desire to renew, to re-shape, to invent, to initiate, to transform, to improve, to undertake, to start anew.

As I pointed out in my previous writings, these tendencies belong to the creative urge *per se.* And yet, in the previous stages of Occidental culture, the focus was on nature, as the ground for man and his life-world, and creation aimed at depicting it, mirroring it, imitating it. With the advent of the modern age which came to its culmination in our century, steps were prepared for the turn toward *man's own forces and powers.* This is precisely the period in which *these steps brought about a radical change in the creative vision itself* which the human being held about his universe of life that marks the advent of a new era. The focus of the creative vision of man with respect to the human province of existence is now deliberately on *his own powers and forces,* that is, *upon the forces, energies, mechanisms, and mysteries of the great fluctuations in the universal schema* of life so that he may penetrate and command them.

This is the creative vision of a new Promethean man, who steadfastly proceeds with stealing from the gods the secrets of the governance of the life of the cosmos.

However, Prometheus was judged by men of antiquity for a blasphemous deed. He obtained the fire but was punished in chains. Contemporary man, on the contrary, has, in his creative vision, freed himself from the chains of being a bound impotent onlooker in the great game of life; he has entered it and he feels free to move in every direction; nothing seems forbidden or closed to his searching eye.

He feels free because he discovered the joints between his marveling eye and his undertaking hands. He not only gazes with reverence upon atoms as the last components of matter, but he breaks them in their actual existence into particles, reaches into their nuclei, discovers their constructive proficiencies, and attempts to recompose them according to his own need, desire, and fancy. His creative vision travels immeasurable distances through space or goes back in time reaching in retrospection the origins and the evolutive steps of life through millions of years. While philosophers used up their lives in past centuries as well as our own to scrutinize the arcane nature of space and time, the scientist seems to have entered into their workings conquering them to expand human freewheeling in the life-network of the cosmos. Human imagination has broken all the barriers of the horizon set up by past cultures; it sees no bounds. As artist, man does not stop at imitating nature, he discovers the very laws of art himself, and instead of adhering to one or other school or style, his creative vision roams in the realm of the laws of his artistry *per se*.

But how can man, whose main effort, whose ever repeated existential effort is to establish his relatively stable dwelling upon mother earth at first, apprehend, understand, or simply face such an outburst of new seemingly endless, ever expanding perspectives on the very ground on which he feebly attempts to place his meager baggage and open his shaky tent? This truly creative vision of the life of the cosmos, of Nature, of a system of life which does not just open a new horizon but which, on the contrary, radiates innumerable perspectives that challenge any attempt at holding things still for more than an instant to have any fixed horizon, is it not a vision for giants, giants of titanic mind who will conquer space, time, all the spheres, and light itself? This creative vision is not a static vision of one creative mind but all human vision of universal life, an explosion of dynamic perspectives before our eyes.

No wonder that philosophers, who used to fashion sand miniature models to show the way in which the mind may mirror within one restricted perspective the structure of the vibrating life of the cosmos (which they have stripped to its skeleton), while tantalized by this outburst, actually shrink into themselves before it, decrying the end of their era, losing faith in the very intellect that before guided them down the trodden path or deserting unconditionally their task!

Facing the unlimited vistas within and without which ever advancing science and technique brings before him, the human being has been caught unawares by these perspectives which throw him out of his complacency with the formerly assumed bonds of nature. Having inherited the habits of a slave who dreams only of shaking off his chains, by which he feels securely attached to his bearings, he now, once they are loosed and seemingly unlimited perspectives open before him, feels his lack of a foothold. How comfortable were the sclerosed schemas of millenial preconceptions, dogmatic assumptions, the inheritance of ontology and metaphysics which situated him within a framework assumed to be lasting, and within which all moves were cautious. Before this gigantic vision of the cosmos, of natural and cultural possibilities, man is overpowered, and it is no wonder that the alarm is sounded over the "death of man," the "end of metaphysics," the "rejection of foundations." The analyzing or speculating contemporary philosopher, groping and lost in the pedestrian scrutiny of his expression (language) level, trails far behind what the light of the conjectural keeps uncovering through unprecedented creative vision.

We need to rise and to assume this vision instead of groping on murky ground, and to return to the goal of determining ultimate principles; but, being at last in the position to approach within a *full-fledged view of nature and of the origins of life* the network of decisive factors in its full extent, *we philosophers are called on to investigate these principles anew at the vertiginous level of "origination and corruption" that the Great Philosopher barely tackled.* We witness not the end of metaphysics but its rebirth, and not the death of God but a novel enlightenment of the mind and the heart. We have this undeniable privilege to have behind us the three periods of Occidental culture with their specific turns of thought — a treasury of instruction and inspiration — as we go beyond them toward a truly definitive philosophical awakening with the light of human creation as our guide. We thus bear the responsibility to ready the new era of culture beckoning at the far-off horizon.

We have thus far offered a succinct philosophical analysis of the element of light in its *primeval* as well as *subliminal* nature. It takes, in fact, precedence over all other elemental forces because it plays a crucial role in promoting the Logos of Life in the constructive unfurling of life. It participates essentially, as we have shown, in the origination of human experience as such.

Now, it is time to attempt an overview of the *spread of human life experiences as such*; in their nature, nuances, propensities, culture takes off and flourishes while human life is carried on. The element of the sea offers us a *panorama of life experience*, and we will study it through the literature which takes great pains to present and deal with it as it fulfills its role. We will discover by this occasion the great passions which animate the human being and their subliminal sources.

CHAPTER THREE

THE ELEMENTAL PASSIONS OF THE SEA

INTRODUCTION: THE PANORAMA OF HUMAN EXPERIENCE AND THE ELEMENTAL PASSIONS OF THE SEA

The element of the sea offers us the panorama of human experience and human virtualities. This drama, in fact presents the entire range of the virtualities of the human condition: first, the spread of human passionate involvements in life; second, the differentiation between the natural promptings of the empirical soul and the subliminal offspring which the soul joins to stirrings endowing them with the threefold meaningfulness that the human condition comprehends, *aesthetic, moral,* and *intelligible* sense. Thus under the aegis of imaginatio creatrix the subliminal, or *elemental passions* endow life with human significance.

The sea as an *element* not only quickens the human drama but also prompts the unfolding of it so that the entire expansion of the virtualities of the Human Condition is displayed: its rhythmic cadences, the play of its intensities as it moves toward its climactic movements. This quickening is due to the revelatory nudity of the deepest, otherwise hidden commitments of the protagonists. It seems the nature of the element to prompt the unfolding of the plot and lead to its conclusion.

Still, the human experiential situation on the sea is strictly correlated with that which the sailor or voyager left upon the shore. For we bring onboard with ourselves our unavoidable baggage. The sea voyage as such marks an existential interval between two phases of regular life. The most striking feature of this panorama is that not only does withdrawal to the sea appear to be an escape from the oppressing business of shore life but that in reverse it is in the isolated situation of the sea that life on shore appears in its most striking and spelled out fashion. Thus, what we discover on the waves about the virtualities of the human condition is of universal validity.

In Melville's dramas of the sea, *Moby Dick* and *Billy Budd*, we see brought to light the natural life forces that are alien to the individual but which he encounters at every turn of his life-struggle. These are

A-T. Tymieniecka (ed.), Analecta Husserliana, Vol. XXVIII, 101–141.

forces of which he is not aware and which he does not understand and consequently cannot conjure up and make a pact with. These forces appear in the sequences of our life-situations no matter what our path; they are forces to which we can either submit as being inevitable robbers of our freedom, our strength, and our aspirations, or which we can fight blindly in order to wrest from them some futile and short-lived satisfaction. These forces which the extraordinary setting of sea life brings out are first, the congenital human pettiness unredeemed as yet by the soul's designs (as seen in *Moby Dick*); second, the crude rule of facts on the sea which severs the innumerable interconnections of life (as seen in *Billy Budd*), and third, the unpredictable vengeance that mirrors the unpredictability of the element (as seen in *Lord Jim*).

But the dramatic situation which the element of the sea creates also brings to light the deep-seated virtualities of the human soul which operate in the subliminal forge of the soul and bring forth a rich fabric of moral significance which pulls life-experience in an opposite direction bringing something to remedy the unmerited sufferings and bring about reconciliation with our fellow men and our selves at a level of freedom above demeaning submission to a blind fate.

With all this in mind we will now, in a gathering of insights gained from several literary works that present the drama of the sea, seek to outline the basic pattern of the virtualities of the Human Condition as they emerge in the play of the human passions, "all too human" as Nietzsche says, being oriented toward survival alone, and the *subliminal passions* which the soul releases out of its own resources, resources of the spirit that tend to counteract the blindness of man's submission to the demands of the natural struggle for life-preservation. The subliminal passions bring into the panorama of experience openings for freedom, for the exercise of the specific virtualities of the human condition which allow man to be the master of his destiny. Of course, we do not suggest that the spread of these two opposite tendencies would lead to disaster at some radical limit. Within the play of associations, fusions, reflections, colorings, tunings which continually goes on in the subliminal workshop in which the brute vital strivings acquire individual-and-societal life promoting propensities, the expanding passions enter into such an extensive and diversified mixture that their subtle distinctions and nuances cannot be sharply divided along any axis. However, there are clear lines along which they fall into groups or categories that we may abstract. Upon firm ground we can recognize distinctive orientations that dominant feelings indicate and can differentiate certain patterns of experience as such into which the complexes of experienced passions fall in their coalescing.

We have already spoken of the category of aesthetic experience which the element of the sea prompts. It is an experience of horizons expanding towards infinity, of detachment from the everyday trivialities of existence, of serenity and repose. Poets through the ages have been evoking these lyrical effects of the sea — elaborating on its beauty, the elevation of the spirit, the sublimeness — which allow man to enjoy this interval as short or as long as it may be between the two phases of his routine shore life with all its worries and cares that drag man down to earth and hide from him the open skies of the infinite maritime horizons.

Yet this clearly poetic elevation that the sea offers for our meditative enjoyment extends to the panorama of human experience that is its counterpart. The sea drama encapsulates human existence; here is a spectrum of other patterns besides the purely aesthetic one. We will focus here only on those patterns of life experience which appear most strikingly in sea novels: (1) the overpowering force of human pettiness; (2) moral man in revolt, the human moral passions on the rebound; (3) the helplessness of the individual in the machinery of the human community; (4) man's ignorance of himself; (5) man's instinctive conformism; (6) the human being's capacity for inward moral transformation within the subliminal workshop of experience; (7) the specifically subliminal virtualities of the human condition — generosity, nobility of heart, benevolence.

I

(a) *The Irresistible Force of Human Pettiness*

A sense of the inexorable force of human pettiness in all its terribleness pervades Melville's masterpiece, *Moby Dick or the White Whale.* We are introduced into this sea-drama by a telling of the preparations for the voyage of the Pequod on shore. And we meet there the pattern of "natural" human stirrings which spring forth from the psyche or the complex of the soul's natural pulsations, sensations, feelings, etc. in their natural responses to the demands of current life. Already in the preparations for the voyage, we encounter at work the most primitive and pervasive vital tendencies of the human being: the shipping captain Bildad accommodates his greedy cheating of naive sailors out of their pay to a rigidly devout Quaker religiosity; with pious deceit he preaches the ten commandments and exults in, his "good fortune." The co-owner of the ship, Captain Peleg, in contrast, takes self-righteous satisfaction in his exercise of the decency of common sense. There is too the self-deceit of Sister Charity, an investor in the whaling voyage, who believes she practices Christian generosity participating by in the last preparations and offering small courteous gestures to the crew. Superstition and malice appear when a chance acquaintance, Elijah, tells the sailor-narrator ominous malicious rumors about the captain chosen for the voyage.

The preparations completed, the ship ready to sail, and those who were outfitting it bidding farewell and returning to their everyday business, the ship under sail, we might think that we part with all that those left on the shore personify: greed, envy, deceit, hypocrisy, self-righteousness, self-deceit, superstition. On the contrary, we encounter the same humanness — but a thousandfold intensified — on board. On board, within a starkly simplified network of action, these attitudes and their motives also, which are usually obscured by a multiplicity of entanglements, come sharply to light. These attitudes toward that which appears to man to be blind forces blocking his freedom of movement are in themselves elemental measures of equipoise, measures that the individual has to invent in order to maintain his life-course.

It is interesting in light of our thesis of the subliminal origin of the specifically human, "higher" or spiritual significance of life — the origin, in turn, of what we call "culture," to note that the two great novelists of the sea, Melville and Conrad, set their human dramas against the background of a reflection on culture's phases of development. Thus, in the drama of the hunt for the White Whale, we become acquainted while still on shore with Queequeg, one of the three harpooners on the Pequod — in whose hands the crucial combat lies — all of whom are men of Nature or savages. Queequeg has the pristine human beauty of a primitive barbarian, an "Indian of royal descent," an "adamic man," and he and his fellow barbarians — pagans — Tashtego and Daggoo, stand by attitudes which strikingly contrast with the pettiness which civilization epitomizes. They are unconcerned with the issues that preoccupy the rest of the crew and display an indifference to or lack of concern with the things which attract the attention of and are endlessly dwelt upon by others. Theirs is an indomitable courage because as real children of nature they take peril and the provocation of danger as part of life; they are as generous in their enthusiasm for danger as they are for life. They stand apart from civilization and do not fall under the category of "moral man," which comes with culture as such.

(b) *The Moral Man and the Life Situation*

As said, the individual not only brings himself on board along with the baggage of his own unchanged nature, but the tiny community that constitutes itself on a ship is organized in a way analogous to the societal organization onshore. The master/slave relationship, to use Hegel's paradigm, is manifested in the societal network on board — first, in the division of the commanding ranks from the subordinate ones. At the top there is the captain; then come the three mates with lesser degrees of authority; and then there is the crew over which they reign. But there is further differentiation as seen in the different treatment accorded at the table to the harpooners, whose valued skills are at the heart of the business of whaling. It is with reference to these societal interdependencies and within this societal framework that the moral conduct of man is shaped and, here, is revealed the propensities of the human being at work.

The narrator, Ishmael, who we first encounter in a Nantucket inn seeking for adventure by shipping himself out on a whaling expedition, is himself one of the crew. We see then through his lenses the life situation on the Pequod, a regular whaling ship rented by the well-known and experienced Captain Ahab and manned by sailors, harpooners, and mates who have come together by coincidence to form a crew. We set out with an understanding of the common purpose of such an expedition, and gradually come to realize the unusual aim which this expedition unexpectedly reveals itself to have, the purpose of the captain.

(c) *The Pervading Moral Pettiness*

The men of the crew, like the people on shore, are in the main a civilized but common sort of crowd. They seem to naturally accept the dangerous career of whaling that is theirs with a basic life attitude of submission. In their brief stays ashore, on meeting at inns and pubs, they revel in their adventures, pass on other tales which they have heard in wonderment (and richly embroidered), and lament those who have been taken by the catastrophes common to their profession — the perils of which they take for granted. They just try to cope with what the turns of life bring; unreflective, they are only half-aware that disaster at sea can be their lot as well. They engage for one whale hunt after another, hoping for the best.

And with the same unreflective submission, they bear harshness, brutality, unfairness, cruelty, revenge, malice, etc. reacting to orders and to the treatment which they receive on board with primitive reflexes; it all belongs to the fare of life.

However, among the crowd there stand out individuals endowed with a moral awareness of right and wrong, personal dignity, self-respect, and higher moral standards of conduct among men. During the voyage, the crew of the Pequod swells with waves of expectation, indifference, anxiety, mistrust; they now fear their captain, the mysterious Captain Ahab — whose missing leg makes him the more enigmatic giving rise to conjectures — and now they give him their frenzied enthusiasm in the heat of battle with the White Whale. But the three mates have an attitude that shows awareness of the general situation of the Pequod, its purpose and chances, the strange character of the captain, etc. They also react to life's provocations with moral concern and appeal to principles of conduct. Provoked at the start of the voyage, they show then the specific personal attitude which they maintain throughout. This attitude represents two different ways in which the moral man deals with provocation and abuse of his moral rights by others.

The first mate, Starbuck, "was an honest upright man," an earnest man of moral strength, who physically is perfectly prepared to endure misadventures for long periods. He is characterized as a steadfast man, one "uncommonly conscientious for a seaman" and "endued with a deep natural reverence." He, however, had during his perilous voyages developed a sort of superstitiousness, or rather, had become prone to "inward presentiments" which were nourished by the fates of his father and brother who had lost their lives whaling. He had taken to whaling not for the sake of perilous adventure but as a way of gaining his livelihood. His honesty proves itself in his profession. He insists that his crew know the peril, that they show courage and not just bravura by understanding exactly the danger and fearing it. His own courage is then his inner fortitude to confront the risks in the best prepared and most careful fashion. This inner fortitude — the fruit of his moral standards and nature — is, as it usually is in life, put to the test. What can he do when confronted with a radical case of business fraud, as we would call it?

The voyage of the Pequod was understood by everybody from the shipping captains who outfitted the vessel down to the crew, to be an ordinary whale hunt. However, a few days after they had set course, Captain Ahab calls the entire crew to the quarterdeck — a call reserved for extraordinary circumstances — to announce that it is not usual whaling that they are going to pursue; they are going to hunt *one* specific individual whale, a whale which is known to sailors around the world and one which has evaded all encounters with the deadly harpoon. A white whale called by sailors around the world "Moby Dick" is the only object of the Pequod's expedition. Taken early by surprise, half-amazed, half-apprehensive, the crew responds with eagerness to Captain Ahab's fierce eagerness in explaining to them the aim of their trip and describing its target.

Not so, the three mates, Starbuck, Stubb, and Flask, who are taken aback by this disclosure. Starbuck, struck by an idea, remonstrated that it was not Moby Dick that took off Ahab's leg. To which Ahab with a violent cry responds that it was precisely Moby Dick who had "dismasted him," declaring that he will chase him "round Good Hope, and round the Horn, and round the Norway Maelstrom, and round perdition's flames. . . ."[17] To chase the White Whale to its death is the aim of the voyage that will lead them to both sides of the globe and "over all sides of earth," as if the white whale stands for all whales, a prototype of whale. The wild mien of Ahab, his violent determination inspires in the men who shipped with him a mixture of ardor and fear. Starbuck, however, confronts his captain on principle. The men of the crew, though astounded by the true aim of their voyage, do not bother to question the legitimacy of this turn of things, being used to blindly following a man. They are seized by a fearful amazement, not far from and approaching admiration, for this strange captain; for it is their universal lot to accept life as it comes provided the wages are right and the life-situation — food, drink, company, and a berth for a body's rest — are as usual.

(d) *The Moral Man in Revolt Submitting to Pettiness*

But Starbuck, the conscientious man, confronts his captain head on: "I came here to hunt whales, not my commander's vengeance," says he, pointing out that vengeance is of no practical value; ". . . it will not fetch much at the Nantucket market." He is outraged on all accounts, not only because the shipping captains and the crew were monstrously misled and not only because to hunt one whale is a scandalously unprofitable, useless task, but foremostly because to seek vengeance — if vengeance it is — on a "dumb brute" which has done Ahab harm not out of any malice but from "the blindest of instincts" is simply madness. Lastly, "to be enraged with a dumb thing" is to the well-balanced judgement of Starbuck, "blasphemous."[18]

In this confrontation based on moral attitudes, the stands of both men are clarified and the stature of Ahab, his standing above common morality, is disclosed. The moral man in revolt is a commanding, independent, highly reflective personality. The ship's principals back at the port and the crew are ordinary, responsible, moral folk, or even men of exceptional intellectual, reflective, and civilized molding. But the recognition of the all pervading forces of evil and their imaginative shaping into evil itself, occurs only when a fully developed moral consciousness is coupled with independence of mind. With the realization of all wrong caused by human pettiness, comes awareness of human evil and in turn insight into its overwhelming insidiousness and the inscrutability of its elusive motivations and a tragic sense of its unavoidable consequences. Evil's unpredictability and constant reoccurrence in varying disguises forms a barrier against which the human being of undertaking spirit painfully butts. Within the soul's subliminal workshop, human pettiness, human evil, turns into evil as such, a fate, a pursuing doom. Its taking shape within the human soul engenders a comparable passion to revolt, a passion to seek evil out, to meet it and break it, a passion which takes possession of the entire person.

(e) *Trust in the Soul Who Rises above the Common*

Ahab is of this superior human cast. He enjoys a strange fame on shore which has brought rumors of all sorts on board, inspiring fantastic interpretations of each tiniest move he makes and keeping the ranks in awe. Not only does he stand out among his crew and direct subordinates by the extremity of his purpose, but his personality manifests his "rare perception" and he espouses concentration on a radical, elevated, superhuman task, no less than to restore justice to creation by eradicating evil at its source. The characters of his companions and the entire spread of life experience come now to light.

When then Starbuck protests his seeking vengeance on a dumb brute, Ahab explodes with a confession of his deepest self. It is not vengeance in the ordinary sense that he seeks. Visible objects are for him but "pasteboard masks"; they have a deeper layer. In each event, in each living act there is what lies behind the molding. The mask is the wall against which we have to strike if we want to free ourselves from the reign of malice which seems to be hidden behind. "How can the prisoner reach outside except by thrusting through the wall?"[19] To Ahab the White Whale is that wall. Although he is not certain that there is a reasoning factor there, yet for him there is an inscrutable malice hiding behind that wall. An inscrutable, incomprehensible, ungraspable, and yet outrageously strong malice. He revolts against this reign of hidden malice with an all-consuming hatred. Taking upon himself the task of defying and fighting this universal evil, he is ready to strike at whatever it is, identifying this combat with the quest for truth.

Thus he defends himself against the charge of blasphemy: he rejects the submission that is the common lot of men. He will not acquiesce blindly to whatever presiding factors there may be, for it is "fair play" that should be "presiding over all creation" and not the apparent envy now at work against us. "Who is over me? Truth has no confines."

In this basic confrontation the natures and moral attitudes of both the first mate and the captain come to light. To the serious, steadfast, conscientious and law-abiding Starbuck, his captain's revolt against the world-order, against its very principle, his daring to question what he considers for better or worse to be definitively settled by a higher instance that he absolutely accepts appears to be demonic, absurd madness. At first rebellious and now dumbfounded, the first mate is overcome by a dread. He does not approve his captain's right to make such a determination. All his inner self is still in revolt against such an incomprehensible — to his simple worldly morality, monstrous — attitude; and yet even though stubbornly resisting his castigation by the captain, he can not but submit with downcast eyes. Upon the open sea, and with the crew in Ahab's hands, he is helpless. At a later time, when the hunt for the White Whale becomes gloomy as it is pursued from the

Indian Ocean to the Atlantic always slipping away and as stories told by sailors on ships encountered make the hunt seem hopeless, then gossip, rumor, and foreboding are intensified a thousandfold by superstition and premonitions, and Starbuck sees an ominous fatality hanging over the vessel. It is then that the temptation presents itself to shoot Ahab.

And so the voyage acquires a definitive orientation. All having become aware of its real purpose, there is no limit to fancy, curiosity, inventive ignorance, and malice in prying into the details of the captain's person, life, and aim and in pondering the White Whale in its strange whiteness and elusiveness. And so waves of superstitious fear sweep over the otherwise brave sailors. The men cannot but be taken by the contrast with the captain of a whaling expedition's usual concern for profit. Ahab's impetuous search intensifies day by day amazing his men whose dread heightens by fact that he has surreptitiously picked out his very own strange-looking, uncommunicative, and absolutely servile crew in order to chase the whale. He repeatedly avers that he must kill the monster himself. This uncharted, wild hunt, one oriented only by hearsay from sailors on ships encountered, moves upon a sea become an "everlasting terra incognita." Ahab himself maintains the watch for the whale, having himself hoisted above the quarterdeck so he can look out from his perch day and night.

Starbuck now feels in his deepest self that this lunatic hunt for a whale who has come to take on in the imagination of the crew the dimensions of an overwhelming monster, a hunt that will go where sailors on passing ships only think that they might have seen him, a hunt with all the frenzied determination of the devil himself cannot but end in a complete disaster for all. Given the tight discipline of a ship, he overtly submits to his commander, and yet inside he is devoured by horror and fear. When a suitable occasion presents itself, he is seriously tempted to kill Ahab as the only possible means of escape from this mad situation. He finds himself facing this dilemma: either to let the ship be led by Ahab to a fatal end, or to murder him. "Wrestling with the angel," Starbuck's law-abiding moral convictions win out. It would need a higher order of moral fortitude to murder Ahab. He renounces any claim to it.

Now, in the second mate, Stubb, we find an extremely different disposition and mentality. Stubb is also a moral man with a feeling for right and wrong in human affairs, for self-respect, and dignity. He has, however, a temperament worlds apart from that of the first mate. He takes life to be a constant game in which one has to be good and cheerful. This in contrast to the serious and conscientious Starbuck who maintains his conviction of impending doom and, while feeling responsible, also feels incapable of averting it as well as incapable of coming to terms with it. Stubb is abusively insulted by his commander in the confrontation with him — an insult to which he gives the powerful and demeaning form of being kicked — and he cannot find rest until he comes to terms with it and with himself. The pettiness of Starbuck's submission to evil, the pettiness of indecision and moral remorse, is here matched by another attitude, namely, that of a "worldly wisdom." Stubb has to submit because the life-situation demands caution, prudence, carefulness. Simply, a revolt would not bode him well. But he has to find for himself a better excuse for this caution. In the supposed dream which he recounts to his colleague, even as he emphasizes the severity of an insult as great as being kicked, he softens this severity by attributing the kick to Ahab's wooden, and not his healthy, leg. Such an accommodating cover-up of his burning shame at being kicked he calls "wisdom." Yes, indeed, it is the wisdom of life's morality to condone accomodation of injuries and insults inflicted by others, to gag natural outrage and accept the pain peacefully.

We either know that things are wrong, but submit to them because of a weak conscience and lack of fortitude, or while recognizing and hurting deeply because of a wrong, we camouflage its sting for the sake of submission because we recognize that revolt does not lead in practice anywhere.

Where Stubb takes things playfully and with humor, Starbuck takes them as a "solemn sight and an omen." Ahab himself observes: "Ye two are the opposite poles of one thing; Starbuck is Stubb reversed, and Stubb is Starbuck, and ye two are all mankind; and Ahab stands alone among the millions of the peopled earth, nor gods, nor men his neighbors!"

Yes, indeed, coming now to to Ahab, we see the nature of his strangeness.

(f) *The Pervading Pettiness of Human Evil*

Can we hold the whale, an animal whose behavior does not enter into the intricate network of human moral/immoral patterns of experience responsible for the terrible irremediable injury he inflicted upon Ahab?

How unconcerned he is, in fact, to causing harm and how indifferent he appears to be to the effects of his self-defense in the chase. The whale as an animal representing nature does not — just like storms, earthquakes, floods, etc. — have any of the motivations, intentions, tendencies which would make him responsible and deserving of punishment. How can we hold storms, earthquakes, floods to wilfully devastate human work? And yet, Ahab invests upon the culprit in his injury all the moral qualities of specifically human evil: malice, deceit, cruelty, perverse dealing, treachery. . . . The irremediability of his injury stirs his moral conscience to its bowels and makes him imaginatively transpose all the types of petty moral evil upon its cause, which then is magnified into a personification of *elemental evil.*

In the chase after this elusive whale who appears in its elemental, strange, almost transparent whiteness at an unforeseen moment and then always vanishes unharmed by the encounter, we have a magnified image of the evil of pettiness. Ahab, acutely aware of the unavoidable weight of human pettiness, wants neither to submit to the pervading force of human evil nor to make any sort of compromise. Finally, it is not even a counter-attitude of revenge that motivates him. He wants to confront these forces in the open and defeat them; he wants to break the invisible barrier to the freedom of man which they represent. The sea, promises to be an open field on which these forces might visibly appear, but in chase after chase conducted by the indomitable Ahab, the whale evades an open fight. In the final chase, the encounter leads to Ahab's defeat and total disaster. The forces of petty evil cannot be defeated or broken definitively. They belong to the cornerstones of human experience and life-progress even as we crystallize the virtualities of the human condition in order to establish our own significance of life.

These forces of the petty evil of humankind are coeval with the specifically human significance of life that civilization installs and are, in fact, an unavoidable constraint upon whatever we endeavor; they condition the manipulation of our life-course through the "force of circumstances"; they so insidiously affect our entire experiential system that we are not clearly aware of them and they elude our clear assessment of them in either the motivations of others or our own. Simply, our self-righteousness sustains all manner of mystifications we create for ourselves.

These elemental forces constraining our human life-course belong to the pattern of human experience. The ordinary type of man submits to injury or seeks to make an instantaneous truce; he hopes to outwit his persecutors this way in some respect. However, each instance of submission or compromise is merely a momentary truce in a continuous assault and holds out no prospect of a definitive solution.

It takes the superior passion of an Ahab to declare war on the intimidating power of evil, to no longer be inhibited in the exercise of freedom, to undertake a relentless chase after evil wherever it surfaces. The whale did not seek a fight and eluded his attacks, but not before destroying him in an as if inadvertant, indifferent fashion.

Must man either submit in some or other way to these hidden forces of human nature, or must he quixotically wage hopeless war on them?

II. THE ELEMENTAL AFFECTS OF THE HEART

(a) *The Heroic Passion*

Had anyone proclaimed it his life-aim to fight universal human evil wherever it might appear and however unexpectedly and to fight it until it is definitively destroyed before any company on shore, he would have certainly been met with bewilderment, incredulity, and a subconscious reaction questioning his good sense. People do learn with their adulthood that human evil is ever-present and ever-persistent in its pettiness. Moral people do accept it as a fact of life and accept the pettiness, that is, the pious hypocrisy with which it is coated to such a degree that we are ultimately misled about its true nature, coming to take the hypocrisy at face value. And so malice is excused by imputing other motives to it, and what is vengeance is attributed to other reasons, and theft becomes another way of allocating property, etc. No one with good sense will bother to seek the true reasons for wrongdoing and pursue it in a court of law unless there are some special circumstances. If he does so, he might even gain and have justice restored to him, but he also must be prepared to be further abused by clever lawyers and to be mistreated by his friends as one who seeks trouble, a troublemaker simply.

For people seek to live quietly, above all, and to avoid, whenever possible, trouble, provocation, etc. Thus, it takes a person not exactly in, his right "right mind," as folks say, to not only denounce the pettiness of life's way of operating as evil but to recognize its universal pervasiveness, its inscrutable reasons, ways, disguises, elusiveness, etc.

This really means to set oneself against the wind. Lastly, to challenge the justice of the tolerance of evil in life events, to challenge even the ultimate forces who decide to tolerate it is in the eyes of worldly morality an outrage, a blasphemous act of questioning. For the tolerance of evil gives to all in the last instance, dispensation of consolation and comfort in an inscrutable fashion. In its mysterious nature, the source of this goodness is hidden behind the walls of reality; it is the source of a hope, as petty as the evil it fights, that the worst of destinies will not be ours.

Thus, Ahab stands out from common mortals as a "grand, ungodly, god-like man" . . . above the common.[20]

(b) *The Passional Complex of Fortitude versus the "All Too Human"*

We must ask how it is that a man like Captain Ahab can tower over common humanity, penetrate the nature of the ways of the world, apprehend that their pervasive universal nature is a feature of human reality, and, finally, take it upon himself to destroy them singlehandedly. When we discover the demonic passion of Ahab to destroy the perfidious moral order of human affairs, we must, along with previous interpreters of Ahab, wonder whether he is not destroyed by his recognition of the universality of evil, his resentment and hatred of it. Do we not deal here with a brute instinct taking the shape of the devil himself? Does Ahab's unique intention to make a panacea out of his hatred and vengefulness not devour his heart and his humanity?

What sort of man must Ahab be? How is his awareness of evil reflected in his being so that he can call up from within himself this singlemindedness, this fervor, this all-consuming determination? Lastly, we must wonder whether is it not precisely by the fact that a man like Ahab can rise above the pettiness of the crowd that this pettiness comes to light as such?

By all accounts, singlemindedness requires heroic passion, a unique fortitude of soul, to sustain its determination to accomplish the goal it sets for itself, and — contradicting what we have just surmised about Ahab — along with that, a generous endowment of the purest human solicitude and compassion for humanity.

(c) *The Passion of Heroic Courage*

We have mentioned the "superior perception" of things that undertaking this quest presupposes; more than that a courage superior to the drives oriented to common life-interests must be possessed so that one may disentangle oneself from the usual life-sustaining network of preoccupations that are shared with the crowd at large. The sea symbolizes for Ahab the realm of freedom from quotidian shore involvements, freedom which he has wrung from within himself allowing him to see his purpose directly and clearly. Melville shares with us "that mortally intolerable truth; that all deep, earnest thinking is but the intrepid effort of the soul to keep the open independence of her sea; while the wildest winds of heaven and earth conspire to cast her on the treacherous slavish shore . . . ."

On the shore, in the port, there is the danger of peaceful slumber and forgetfulness of the great concerns. The port is "pityful; in the port is safety, comfort, hearthstone, supper, warm blankets, friends, all that is kind to mortalities." What courage a man has to muster not only to tear himself from this comfort but to avoid it and to flee upon the heartless, indifferent, immeasurable, and unforeseeably perilous field of the sea embarking on a lonely voyage with a single purpose in mind! The ship is a fragile vehicle, not even his own but rented under a false pretext and manned by people who were unaware of his purpose and whose allegiance he has not only to win but to sustain. The courage to tear himself from the natural life-world and to establish for himself a new field of life nourished by his mind, one in which he has to take all responsibility, indeed, everything, upon himself, while being understood only vaguely at best, is COURAGE which only the most vertiginous deployment of the subliminal soul can generate; this is the courage of all heroes, all revolutionaries, reformers, protectors, pioneers, saints, geniuses. . . .

(d) *The Pure Solicitude of Heart That Shames Feeble Pity*

Moreover, in this specific case in which heroic courage is directed against the devil himself, he who declares war upon evil must have first known and penetrated it. And to understand it, he must somewhat identify with it. This necessarily means the blighting of the serene, pacific, harmonious, and trusting state of mind to which every average mortal aspires. But let us listen. AHAB SPEAKS:

> Oh! time was, when as the sunrise nobly spurred me, so the sunset soothed. No more. This lovely light, it lights not me; all loveliness anguishes me, since I can ne'er enjoy.

Gifted with the high perception, I lack the low, enjoying power; damned, most subtly and most malignantly! damned in the midst of Paradise![21]

Does this mean that all empirical enjoyment is extinct, that he is no longer capable of human feelings and affections?

Moody, with "a crucifixion in his eyes," and with a domineering stature, exuding a sort of "sultanism," far away, however, from any dictatorship, but simply possessed of a superiority of mind, Ahab inspires not only uneasiness and fear in his officers, but doubt in his humanness. Indeed, he seems to be foreign to simple human pity. When implored by the captain of an encountered ship to interrupt his hunt to search for a missing boat in which two of the captain's sons are missing, he steadfastly refuses the entreaties and continues on his own set course. And yet, while all empirical affectivity seems to be inactive in him, when confronted with the total deprecation which his civilized, moral crew, all the white members of which live comfortably convinced of being good Christians, show to the black boy Pip on account of his madness and cowardice and with their complete indifference to the life of this human being when he is about to drown because they believe that he is not worth rescuing, Ahab acts.

Those who beget such an unfortunate creature commonly abandon it and any others who do not succeed, for those who cannot play the game of life are disowned, despised, as well, as useless creatures — less than human. They float at the margin of life and are at the mercy of people. Thus, the Pequod's crew refuses to let Pip climb back on board after he has fallen out of the ship. It is precisely Ahab who then responds to the deepest of miseries. Not easy pity or compassion, but the innermost passion of the soul, that of *human communion*, springs forth in Ahab's being; this is an encounter involving the deepest of ties at the "inmost center" between two human beings denuded of all the petty paraphernalia of worldly life-conventions and hypocritical morality, the tie of the purest of solicitudes. In a subliminal movement of the soul, Ahab recognizes that the black boy Pip is "tied" to him "by cords woven by . . . the heart strings." Taking him by the hand, Ahab takes the homeless creature to his own cabin declaring that it will be Pip's home as long as Ahab lives. This movement is immediately received by the boy who responds: "Ah, now, had poor Pip but felt so kind a thing as this, perhaps he had ne'er been lost! This seems to me, sir, a man-rope; something that weak souls may hold by."[22]

Thus, a subliminal recognition takes place between two human beings as radically different by human worldly standards as they only could be: here, a young bell boy, an "idiot" by all accounts, helpless, weak and life-forlorn, and, there, a powerful, strong educated ship commander. And yet they recognize each other as both being on the verge of existence, existence denuded of all human ties and the comforts they bring. The hapless boy on losing the last strawhold on his existence, physical access to the ship, was facing death. ". . . but Death is only a launching into the region of the strange Untried . . . the first salutation to the possibilities of the immense Remote, the Wild, the Watery, the Unshored. . . ."

That same shoreless voyage lays before Ahab's eyes. These two beings share in their deepest self the innermost pure solicitude of the heart. Ahab is touched at the innermost of his heart by recognition of mankind's obliviousness to universal suffering as illustrated in the fate from which he saves Pip. Gods, goodness, etc. appear meaningless to people caught up in practical activity; they forget all the precepts of moral conduct toward others. Gods appear oblivious to the fate of man. Yet a human being like Pip is "though idiotic, and knowing not what he does, yet full of the sweet things of love and gratitude."[23] The sea is a field on which awareness is heightened of the human situation of being at the verge of existence, of living from one moment to the next; entrusting oneself to a fragile vessel and being irremediably unshored facilitates the revelation of this truth and the laying bare of the deepest zones of the soul.

And yet there is still one major movement of the subliminal soul that is the more easily spurred by stripped down monotonous shipboard existence, especially when on a metaphysical pursuit bound not to port but to a kill.

(e) *The Stirring of Nobility*

Starbuck, who remains in stubborn and radical opposition to his captain's aim and who as a moral man of superior conscientiousness could not kill him not even for the sake of saving the entire crew, has also remained a compassionate being. On board a ship in full sail, when there is a clear sky, a steel blue sky, with a quiet wind, and only a gentle breeze playing on soft waves in the morning sun, there is a corresponding climate of peace, of abandonment to the timeless moment; and then occurs a quiet recapitulation of the inner self before the inquisitive eye of the mind. This is also the psychic climate most propitious for sharing one's innermost in confidence with a fellow man, for co-communicating in trust our innermost with open hearts. More, it is a spiritual climate which allows for the rare move of a noble elevation of the heart, both the lifting of one's self toward trust in the absolute truth of one's being and being able to receive and appreciate this experience.

And so Melville shows us how one such serene morning, so in contrast to the gloomy mood of Ahab, "the lovely aromas in that enchanted air did at last seem to dispel, for a moment, the cankerous thing in his soul." And his mate Starbuck seeing him is moved by his weary, sorrowful figure in such contrast to the serenity in the air, is moved so that it seemed to him that he could hear in his own breast the "measureless sobbing" bubbling in the very depth of Ahab. By a solicitous movement of the heart, he draws near to his captain and remains there motionless.

In a passionate soliloquy Ahab makes his first mate privy to his deepest thoughts and concerns. He confides his wretched feeling of being used; he is heavy with the burden of forty years of the desolate solitude that surrounds "a captain's exclusiveness" of his slavery to his ocean isolation from all nature, of the "dry nourishment of my soul," of his young wife left alone on the shore, and of his maimed condition, the work of a senseless brute of a whale. He shows that he understands the life, family, and shore attachments of his first mate: Ahab sees Starbuck's dream of returning safely home as if it were his very own, and with the noblest generosity, he entreats Starbuck not lower his boat in the chase for the White Whale but to remain safely on board. He alone, Ahab, will risk his life. He alone will give chase to Moby Dick. Starbuck should not take the hazards.

Stricken at the center of the heart by this discovery of not only the full generous humanity of Ahab, but also, and most of all, the man's concern for the lives of his people and his grand move to take all the danger upon himself. Starbuck recognizes the nobility of this gesture within himself — only through a sublime movement in our own soul may we recognize the same in another — and he cries out: "Oh, my Captain! my Captain! noble soul! grand old heart, after all!"[24] In this so elevated understanding, Starbuck immediately entreats the captain to abandon the chase, to give up the obsession with Moby Dick, change course, and return to Nantucket. "Why should one give chase to that hated fish!" Although he is capable of responding to the noble man, nevertheless he obviously does not understand the heroic purpose of his captain's obsession. He cannot penetrate the recesses of the soul where Ahab seeks to understand himself; he cannot see under "whose command" the man is; against all the he naturally cherishes and loves, Ahab relentlessly pursues his receding goal. He does not feel that he is his own master in this undertaking. He could not do otherwise, no more than could a migratory bird seek to return from his winter abode by a different route, no more than could the sun, the moon, or a star plot a different course. If all things seem to be moved by some invisible power, how could he believe that he is his own mover, that he makes his own heart beat? How does it beat, how does he think, "unless God does that beating, does that thinking, does that living, and not I."[25] He is no more capable of reversing his course than all those things whose motion is pre-established are able to change theirs.

And so this profound communication of souls, while allowing a profound life experience of the highest moral order does not suffice to bring about a reconciliation between these two people. Starbuck steals away from his captain in dread and despair.

(Ahab keeps to his decision. During the final chase for the whale, the ship, with Starbuck on board, is ordered to keep at a safe distance from the action. It is Ahab alone who personally attacks Moby Dick. Despite the order, fate decides that the ship will go down, and none but Ishmael survive to tell the tale).

To conclude, our investigation of the life-experience and human virtualities in Melville's sea drama let us say that we might not be able to control our basic life-course any more than the heavenly bodies can theirs; we might have individual confines within which our inner self must move and cannot trespass beyond. But within these confines we are definitively capable of drawing upon the resources of the Human Condition deposited within the human soul to confront the reign of the universal pettiness of life within ourselves and to transmute the subliminal virtualities in a crystallization of our bents, tendencies, and longings into the purest — that is, free of life-subservient self-interested significance — movements of the heart in communication with other human hearts and thus establish with them the spiritual life-significance of a culture.

The creation of human culture is, indeed, at stake here, in a double way. First, there is, as we have just pointed out, its origination within the subliminal workings of the human soul; second, as has already been intimidated by the fascination with levels of culture manifest in Melville's inquiry, it seems that previously acquired culture accounts for the capacity of the human being to advance in his life-experience to the highest realizations and acts of moral experience. We will discuss this in the next section with fresh illustrations from Melville's other sea drama, *Billy Budd*.

III
THE FURTHER RESOURCES OF THE SUBLIMINAL SOUL AND THE ELEMENTAL PASSION TO BE TRUE TO ONESELF

(a) *Generosity of Heart, Benevolence, and Reconciliation with the Inexorable*

Melville's creative vision of the human life situation does not stop at hopeless doom, at the radical revolt against evil forces, the attempt to wage outright war against it, and resulting disaster. To the question of whether all we can do is is either accept evil and submit to it or confront it in the open and fight it, our subliminal resources offer an unexpected answer. In fact Melville probes still deeper into the human soul in his sea tale *Billy Budd*.

We encounter on the Bellipotent a setting similar to that in *Moby Dick*, but two main protagonists of a different cast: Billy Budd, the innocent, handsome sailor, and Captain Vere, a fine specimen of an upright man with strong moral convictions. Captain Vere of the Bellipotent has, as is usual with Melville's commanders, an intellectual and civilized nature. More, he is considered to be an altogether "exceptional man" in the eyes of his friends, "a noble fellow . . . Sir Horatio (meaning him who became Lord Nelson)." But we see also "a pedantic streak running through him," his deepest concern being the probity of his convictions and their application. To complete this picture, Captain Vere is characterized as being little communicative or sociable, not given to the usual comradeship of life on board, "a bookish type." This feature of his character will play an important role in our understanding of the denouement of the drama.

Billy, the handsome sailor, is presented by Melville as an example of a man in whom "the noble descendence" is as evident as in a blood-horse. He is "cast in a mold peculiar to the finest physical examples of an Englishman"; he "shows in the face that humane look of a reposeful good nature." More, he is in his congenital goodness innocent of the world of men with its malice. In this respect, Melville sees him as "little more" than a superior type of barbarian, "as Adam presumably might have been ere at the urbane serpent wriggled himself into his company." In short, we have here a representative of unspoiled pristine nature. Unaware of the malicious tendencies at work in civilized humanity, Billy in his trusting simplicity falls the victim of calculated human malice. Perfectly innocent, he has to pay with his life for the malice of men. Condemned by the upright captain who has to condemn him irremediably because he is a man of duty and the War Code has to be obeyed, Billy can not understand the cruelty of the man who condemns him despite his innocence. The cruelty is inscrutable to him, and he cannot accept submission to it. Do we ultimately know why we have to submit?

Awaiting his execution, we believe with Billy that an atrocious act of unjustifiable violence is about to be perpetrated. As things unfold, we are struck by what, within life-context seemingly governed by networks of human tendencies which represent the promptings which led to this fragrant injustice, appears to be an amazing turnabout. The law has to be respected even if the sentence is perversely unjust. And yet, after an eye to eye and heart to heart encounter with Billy sought by Captain Vere, the condemned comes out with a forgiving heart, serene and reconciled, with all the purity of his innocence. We are not told what transpired in their interview, and yet we know it: to Captain Vere the sentence condemning an innocent man to death must be in all its proportions as terrible and tragic as it is to the victim.

To follow his tragic inner struggle, we must to begin with ask how it is that the otherwise uncommunicative captain who normally would shun a face to face encounter and offering of an explanation with a friend is taking the drastic step of seeking such a session with a young subordinate. He must be strongly motivated to so depart from his usual frame. What stirs him so strongly that he undertakes to confront the desperate young boy? Obviously he is caught in a dilemma between dutiful obedience to the law and the voice of moral conscience denying its validity. The decision clearly has shaken him out of his normal composure; to alleviate the tragedy, he seeks to talk to the victim man to man; he hopes that in showing him his own dilemma, in letting the young sailor see his own personal tragedy at a larger, deeper level of human life than his exclusively subjective perspective, that they may meet at this deeper plane even as they have to, to all appearances, part ways.

Billy Budd in his strength and purity of heart, not knowing the evil of the world and thus not caught up in its coils, is touched in his deepest self. His subliminal forces being released, he can accept suffering from a man who believes in the law as the safeguard of the subordinates entrusted to him — of civilized humanity — and believes that to defy it would mean to reject its validity. A "foundling of a noble race" is able to kindle sympathy and generosity toward a moral man of the law who means no wrong but is a victim of the law himself. There are released in him further passional resources of the soul which strengthen him allowing him to forgive and face the unfair, accidental loss of his life. Billy Budd's reconciliation with his doom comes from a deal made between the two at the very depth of the subliminal workshop of the human being where *benevolence* toward others takes precedence over our own self-interest, and *generosity of heart* allows us to cross barriers, to cut through societal and cultural schemas and individual propensity. It allows for two human beings to meet in the deepest human fellowship. It is in this fellowship that the bitter despair of Billy is redeemed, as well as Captain Vere's deadly remorse over having to decide on the death of an innocent man.

We see here the merciless cruelty of rational morality denounced. And yet — as is brought out in both of the novels by Melville here considered — this morality constitutes the backbone of our civilization. Would not the virtualities of the human condition which introduce the moral sense as the foundation of human interaction through the benevolent/malevolent sentiment (even though benevolence is opted for and the claim to protect man from the malevolent tendencies of others is made) and which are the origin of our moral codes and systems be vitiated and perverted by the undue way in which they are rationally treated as laws and regulations objectively and universally valid for all?

As it appears, the moral law does not protect the innocent against the snares of the malicious. The release of the specifically personal resources of the human soul — subliminal resources — is necessary to bring forth this deepest generosity of heart that redeems what is good, loving, heartfelt, and honest in the human being in its purity and innocence of the world as if it were still the first day of creation. In the subliminal mystery of human brotherhood over against the terrible and unconquerable evil, lies the mystery of reconciling with it and thereby getting the better of it, of how to conquer its destructiveness in our innermost self. We might be obliged to submit to the merciless assaults of human pettiness — Billy Budd, the handsome innocent sailor, is hung — but this cruelty has been redeemed in Billy's heart and this gives a unique significance to his destiny.

We may look at the human experiential situation in the mythical perspective of paradise and man's fall. The freedom of the first man in paradise resides in his innocence. Once the distinction between good and evil emerged, it seems that the situation of human life experience became suspended on the freedom/constraint axis. The human being endowed with moral judgement longs for freedom, and he either revolts against the constraints or submits to them adopting one or another attitude. And yet, even when he revolts and fights for freedom, he is like Ahab defeated — inevitably; he remains irremediably within the bounds of evil's pattern.

Despite this, the Human Condition holds virtualities of the soul, the human soul, the unfolding of which allows man to endow his life with significances through which the very meaning of facts, events, decisions, and human relations is transformed. As we have seen in Billy Budd's tale, the inner resources of the subliminal soul reach further than the empirical resources informed by the moral sense. No doubt, they receive forms and forces which prompt the inventive constructivism specific to the human being and which are in this respect embodied in the societal circuit of life. However, they are then oriented toward this end and remain in the service of the societal interaction in which there is a measure of freedom of choice between good and evil. Yet even this measure of freedom is apt to be used for the vital societal advancement of the human being and thus be vitiated by rational calculations and malicious propensities. The specific resources which the Human Condition reserves for the subliminal soul, however, reach beyond rationalized societal morality with its universal laws and their application. With the specific subliminal synergies of the soul, benevolence, congenital innocence of heart, and generosity may be kindled and may surge as passions which inundate the field and fill the human soul with good will towards all humanity, nature, the world, the All-encompassing. These passions allow the innermost life-redemption, whatever might befall us. They are the basis of our reconciliation with our inevitable doom. They foster the deepest personal communion among people. They lift the significance of our daily life above societal, natural, personal constraints giving us a destiny above the life-course.

(b) *Repentance, Recognition of Faults, Indulgence and Mercy; the Pattern of the Subliminal Passions Bridging the Redeeming Experience and the Hard Facts; Establishing Equipoise within Communal Existence*

The creative vision of Melville's sea panorama of human life experience is not yet closed with the surging of the life-redeeming inner passional resources of the soul. We have still to face the radical cleft (hiatus, abyss) between the redeeming inner forces and the crude reality of facts. Indeed, the inward redemption of the unjust death sentence does not influence its outward effect: the cruel and unjust act has to be carried out, nonetheless.

In Melville's last poem, "The Bridegroom Jack," a new sea tale and one recounted by the protagonist himself, a further deepening and enlargement of the pattern of the subliminal resources of the soul is offered, a conclusion, as it were, of the author's quest for the virtualities of the human condition and of man's control over his life.

To be brief: Captain Turret, commander of the ship, first condemns the mate Finn to be flogged according to the rules of the War Code — the iron, all-powerful law. After seeing the prisoner's painful repentance, he, as if under a new impulse, exclaims, "Submission is enough." After what we have seen of ship discipline in *Moby Dick* and *Billy Budd*, this change of heart represents a most eloquent, entirely novel disposition and attitude: mercy.

Mercy seems to be a unique type of a subliminal passion; yet it draws upon several others. Captain Turret's change of heart obviously took place first out of compassion for the sailor when he saw his contrition and remorse for his misdeed. Yet contrition is in itself already a subliminal passion insofar as it does not belong to the spectrum of direct life-promoting strivings and shunnings; on the contrary, it involves a return of the soul from its outward directed societal and survival tendencies to its deeper self and a coming upon the finer moral stirrings in which self-centeredness is dissolved and a general human conception of life acknowledged. Our innermost commitment is reconsidered, and a passional surging moves us to efface the dead past's effects upon ourselves and to "reform." Thus, the movement of compassion in the captain is in unison with the mate's inward movements. Furthermore, instead of rigidly sticking to the application of the all-powerful code of law, Captain Turret, in his movement of compassion, strikes a flexible cord. By taking in the inner state of the culprit he eases within his conscience the principles governing the application of the rule to the point of liberating from the subliminal forge a concealed *indulgence.* At last, to take a stand on the situation, an *act of mercy* dismisses the crude and fanatic self-righteousness of the legal mind. With the act of mercy we reach the apex of the moral subliminal passions: *magnanimity,* which is *the vortex of the moral significance of life.*

We have seen in this investigation the unique resources of the human soul at work as they proceed from the subliminal workshop of the inner man and are called upon to break through the dimensions of man's natural and societal life-interests to inaugurate a uniquely human spirit — to not only break through human pettiness, but to counteract its factual effects through magnanimity. If the petty malice, calculation, and cruelty of man consign him to the very bottom of the scale of life, magnanimity crowns him and lifts him to the summit.

Melville's vision of humanity is, in fact, suspended upon extreme types; one is the pristine purity and beauty of the barbarian, and the other is the dominating individuality of the superiorly educated, reflective, intelligent and independent mind as found in a man who takes responsibility for others and decides the course of their lives. In between lies the civilized but average crowd. From our discussion it is clear that the surgings of the subliminal movements of the soul are not indifferent to the level of the person's civilization. Civilization seems to have, on the one hand, inherited the full consequences of the disaster of the mythical fall of man; on the other hand, it appears in its highest unfolding to also be the ground for the liberation of man's innermost subliminal virtualities, virtualities which, as we have seen, allow him to equilibrate his life situation, even to avoid its pitfalls.

(c) *The Radical Test Within the Intricate Web of the Human Community*

Joseph Conrad seems in his "human psycho-drama," as his masterpiece *Lord Jim* has aptly been called, to have radicalized the elemental challenge of the sea by discovering the true core of the human self as well as the significance of this challenge when aimed at the singular individual for that individual's role and place within the entire human community.

The sea, in offering man only the absolutely unstable and shaky platform of a boat, brings out the utmost fragility of his entire beingness; it allows the human being to be tried as to his/her strength, valor, reliability in radical either/or, life/death situations that cannot be postponed, evaded, or conjured away. It offers him/her unique opportunities for probing the very core of humanness, that is, the "true" state of moral selfhood.

In Melville's character of Ahab we had an "uncommon man" who challenged radically the petty evil of conventional human life. Ahab, a man of maturity, experience, and strength, set out to directly combat what he believed to be the source of the evil. The chase after the White Whale personalized his fight. However, in this allegoric fabulation, Melville did not enter into the concrete characteristics of the societal establishment. Conrad's hero, the young Jim, second mate on the *Patna*, enters through his uniquely concrete predicament into the heart of societal conventions with not his life but his legitimate role and place within the community of man being at stake.

Thus, dealing in the concrete, Conrad goes farther than Melville in raising the issue of the moral selfhood of a superior man in relation to the human community. First, the relation of this true moral selfhood to the human community is examined within the civilized community which has betrayed Jim, in accord with the societal order and its laws. Second, this moral self is exposed within a primitive human community, and there Jim has the opportunity to unfold and flourish. Still, even there, at the peak of his strength and fulfillment, a small adverse incident is enough to topple his gigantic moral accomplishment, with his death being the penalty paid. The question that emerges in this moral darama is that of what Jim's "true moral self" is with respect to humanness taken in the absolute sense.

In fact, we see in Jim's tale that the domination of the community, of the societal order, over the individual is inescapable from one end of the globe to the other. Throughout the whole of humanity, civilized or barbarian, man's communal "fit" stands out. This domination is based on societal laws that in their nature consist in a petty consensus which will necessarily collide with the moral stand of a person of superior culture.

What is the significance of this collision?

In the Author's Note (June, 1917), Conrad, in a heated reaction to the criticism of a lady who reproached him for Jim's "morbidity," answers that there is nothing morbid "in the acute consciousness of lost honor," adding that "perhaps, my Jim is not a type of wide commonness." However, he is not the fruit of sheer invention either, and his predicament should not be one foreign to the European cultural sensibility. He is not a striking figure of a hero, but the author "saw his form pass by — appealing — significant — under a cloud — perfectly silent. Which is as it should be." And it was up to the author, as he tells us, to find the proper way, "fit words," to show this figure in his peculiar significance, simply because he was not striking outstanding: "He was one of us."

What, however, should we understand under "acute consciousness of lost honor"? To what realities of the soul does Conrad refer, and how do they stand beside other passions and proficiencies of the human psyche? Lastly, just what is the dominant moral passion which carries Jim's relentless effort to redeem his "honorability," to be reconciled with the human community, we will learn from the masterly fabulated tale itself.

Anticipating our inquiry, let us point out, first, that Conrad is here digging into nature itself and into culture and its personal foundation with the human soul caught in the unfolding of its subliminal faculties: imagination, reflection, conscience. Second, drawing on our previous analysis of the passion of generosity as stemming from the deepest assumption of self-respect, I will propose here that we discover in Jim's plight an even deeper ground upon which human personal morality and self-respect is established, namely, *the passion to be true to one's self.*

Here we must pause. Is Jim's conduct still received as if he were "one of us"? In our present day devaluation of the classical virtues, have we not really tacitly traded the prerogatives of feeling "honorable" for pragmatic expediency? His conduct, as the author specifies, is, in any case, not common behavior found at any stage of Occidental culture. Our inquiry into the state of mind which grounds Jim's attitude toward the world and himself will show us that the term "honor" expresses in a summary way the state of the inner moral self which under different names remains the touchstone of Occidental man.

What the significance of Jim's moral stand is with respect to the absolute standard is, at this point in our inquiry, an open question, and may be an unanswerable one — as Conrad himself suggests. Man is a communal being, after all, and the moral propensities of the Human Condition refer to the communal life. However, it is due to the ways in which people like Jim react to the perplexities into which they run within established societal law as it is applied in challenging situations that the law is improved and the moral state of human communities advances. Without such people the societal order would fall into disorder, violence, and barbarism.

Who is Jim the sailor as "one of us" — Conrad intimates Jim's being "one of us" in several respects — and in his true self? These are the questions running through this story of the intimately personal, and communal as well, tragedy of a single major protagonist, a tragedy that ties its dramatic knot and disentangles it within Jim's innermost being as it expresses itself in his moral conduct.

We first meet Jim as an exceptional water clerck who has made himself popular in various Eastern ports for his extraordinary performance of his duties. These duties, the author tells us, are extremely varied and demanding: "To the captain he is faithful like a friend and attentive like a son, with the patience of Job, the unselfish devotion of a woman, and the jollity of a boon companion."[26] The needed combination of these, even taken singly, exceptional qualities makes good water clerk very rare. However, Jim topped all these with the doubt-free, many-sided courage with which he would brave the wind and the waves in order to be serviceable to the ships approaching port — without any emotion, but, to the contrary, an exulting certitude.

With this spirit and "avidity for adventure," Jim joined the crew of the *Patna*, a ship carrying eight hundred Eastern pilgrims to a holy place. Sudden damage to the ship led to a critical incident. On account of this, we meet Jim some weeks thereafter at the police court where an official Inquiry is being conducted. In radical contrast to his character, he stands there as the culprit, a shameful coward who together with the skipper and two engineers deserted the ship and the eight hundred passengers who were entrusted to their care and whose lives were then threatened, a breach of the code of the sea. In the eyes of the court representing the community this act is a matter of clear-cut facts: nothing could justify the abominable breach of trust. As Captain Brierly emphasizes, sailors are "trusted," and it is precisely "that kind of decency" which says stay at one's post that people's trust relies on. More, it is this solidarity with men who entrust lives and/or possessions to them that holds the sailing community together and gives it its dignity before the rest of the world. To this decency sailors commit themselves and their lives absolutely. In Captain Brierly's words, "A man may go pretty near through his whole sea-life without any call to show a stiff upper lip. But when the call comes . . . Aha!"[27]

By the rules of the civilized community, Jim was irremediably condemned. But to Jim there is nothing so clear about his case. He is jolted in his entire being. He wants to assume full responsibility for his act, and significantly he submits himself to standing trial all alone, the skipper and his two other "partners in crime" having cleared out in time. He refuses to avoid the ordeal of the trial even when prompted by Captain Brierly to do so. And yet, he is absolutely revolted by the significance which the court, the code, the public, and the brute facts themselves give to his act. He wants to elucidate at any cost the true meaning of the events which led to it. But the magistrates want to learn the facts and nothing but the facts, "as if facts could explain anything!"[28]

Jim committed an act on the sea which had to be condemned by the code of the human community. And yet those who presided over the Inquiry into the *Patna* incident recognize that he is a "better sort" of man.

(d) *The Unfolding of Moral Tragedy: The Inwardly Held Ideal versus Conduct*

Captain Marlowe, who with all the sailors in port attends this unprecedented trial, is struck by the discrepancy between the disgraceful deed of desertion and the person of the culprit. Jim appears to be a straight, fine fellow: "this frank aspect, the artless smile, the youthful seriousness. He was of the right sort."[29] That he was facing the court "practically of his own free will" was itself a "redeeming" feature. In short, he did not belong to the sort of men who like his skipper and the engineers run away to save their skin. An altogether other sort.

Jim later tells Marlowe that from the time he committed the irreparable act of jumping the *Patna*, he had been in a hebetated state. He submits himself to the judgment of society, hoping that the Inquiry will help him sort things out. He accepts full responsibility for his act as well as the validity of society's code. He is struck, to the marrow of his bones, by the radical discrepancy between his inner state, his attitude toward life, his dispositions, and, lastly, his noble dreams and the act of desertion that he committed. He wants to understand it. Most importantly, his sense of responsibility goes deeper than the laws of the human order; it touches his innermost self. He, in fact, cannot conceive why and how he could have committed such an, in his eyes, incomprehensibly despicable act. How could he have missed what in his eyes was not "a call to show a stiff upper lip" but an extraordinary occasion to measure his forces, to fulfill his dream of heroic dedication, an occasion that happens only — if it does at all — once in a lifetime?

His act seems so foreign to his inner self, to his feeling about himself, that, tantalized by its occurrence, he is prompted to seek its causes. He tries to get at them in his efforts to answer the questions of the magistrate. His drama began not with the crisis on the ship but with his innermost urge to translate those events into a most accurate and detailed account of the facts that in succession led to his desertion. The radical discrepancy which he could not understand makes him feel that behind the empirical facts there must have been an "invisible factor," a directing spirit of perdition. "There was a true horror behind the appalling face of things." In order to find out what that horror consisted in, he tries to conjure a revelation that would indicate what prompted him, and how, to make the decision to leap onto the lifeboat. Before the court he exerts his mind to the utmost to review the minutiae of the facts circling within his mind over and over again in "tighter rounds." But at the moment when he thinks he is quite near and is just about to strike an opening out of the closed circles, an opening that would yield the sought after insight into how he could have committed a deed so contrary to his nature, a deed by which "he cut himself from the rest of his kind,"[30] the magistrate would cut him short by asking for a yes or no answer. The magistrate was not interested in the state of his soul.

The Inquiry, then, fails to respond to Jim's innermost need to explain the reasons for his decision. But was it a decision?

Jim's tragedy unfolds with the failure of the Inquiry — which he decisively stays with until its end — to give him the opportunity to find the explanation he seeks. Although he accepts the verdict, he refuses to be content with the societal significance of the act without any understanding of its innermost personal significance. In order to live, he has to seek the key to this closed circle of facts. He cannot accept society's verdict as an irrevocable branding. He must learn his "true self" in order to measure its relationship to the discrepancy. Lastly, he has to measure his innermost dispositions against the demands that the societal code puts upon man. In short, he sets out to find clues that will allow him to redeem himself, give him a second chance.

Finding in the reflective Captain Marlowe a sounding board or alter ego in his plight, Jim unfurls his inner tragedy. So begins his quest for elucidation. Describing to him the tiniest moments of the observations and feelings which preceded the incident and those of the incident itself, he probes into his entire functioning. Following him, we discover that at the level of his subliminal self articulation was lacking in his decision. Was it even a decision?

When the irreparable and sudden damage to the ship was discovered and the short supply of lifeboats precluded saving the great majority of passengers, the skipper and the engineers, led by the primitive instinct for survival, decided to lower a lifeboat and quietly flee. But Jim, witnessing it, did not fear for his life. He felt a moral disgust for them while they were working on lowering the boat. His moral consciousness was too developed to even consider the idea of abandoning the hundreds of pilgrims to their doom and of escaping. However, while the primitive minds of the skipper and the engineers were focused upon securing the means to save their own lives, Jim with his reflective and imaginative nature was experiencing and envisaging the full extent of the situation. The ship being doomed to sink any instant, his imagination was spurred to visualize its going down and to live imaginatively his own death. With there being nowhere to turn, nothing to hold on to, this imaginative vision spread dread through his entire being, making him incapable of taking a step, paralyzing him. Such was the weight of the dread on his nerves. Then there was the horror of the thought of hundreds of innocent pilgrims unaware that they will any instant be swallowed by the sea. There is the ominous stillness of the ship and the waters. With imagination playing thus upon his subliminal strings, with his nerves paralyzed by dread, overtaken by awe, he needed time to sort things out. All of a sudden the sky darkened and the wind announced the onset of a gale. The call from the lifeboat urging the fourth crewman to jump created a physical prompting. There was no time. Instinctively, unknowingly, without his will or desire, Jim followed the call and jumped onto the boat.

(e) *The Communal Attunement of Man's Making*

In its elemental power, the sea has here only intensified to the utmost and stripped to its bare bones the crudeness of radical situations which all of us may confront in life. We all go through life handling more or less smoothly the demands of current circumstances. If we perform the task assigned to us dutifully and avoid trouble with our neighbors, etc., as the greater number of people manage to do, we are naturally taken to be a realiable cog in the great machine — or a loop in the knit — and so enjoy a part of the trust that the community gives to its members; we have the trust then upon which the life of the community depends. In its life a human community spontaneously develops rules, regulations, standards, principles, values, etc. according to which it progresses constructively. From our earliest youth we are brought up to enter into its moving schema, and the great majority of us naturally fall within its folds.

We are woven into the web of the community of human interests, and through duty, responsibility, habit, the expectations of others, their attitude toward us, and their appreciation, we are sustained by this web and acquire a secure feeling about ourselves. From our place and role in this web we draw a feeling of confidence, of "belonging." Our self-assurance and self-confidence mean that we are aware of our fulfillment of the expectations placed on us. Our security and self-assurance stem from our conviction that we act dutifully. When challenged, we prove our credibility and "good standing" by detailing our proper past performance and citing testimonials to its propriety by our neighbors, employers, friends.

Though hesitant, wondering, dubious about ourselves in our early youth when we set out to find our "identity," we gain strength of conviction and faith in ourselves and in life from finding our place and role within the community. We live from day to day wrapped up in the mantle of the trust that our community places in us, believing ourselves to be naturally sincere, faithful, compassionate, honest, respectful, and decent.

To obey the rules and to avoid friction, confrontation, awkwardness, unpleasantness, and conflict means, however, to remain on the established surface of things, to see only what is obvious to everybody. In conflict situations, however, the deep foundations of these rules, as well as our "true" nature and inclinations, come to light. But ordinarily we continue, feeling complacent about ourselves, our community, and the world. Should contrary events shake our faith in our capacity to deliver what is expected from us, we will be thrown out of the composure indispensable to going on with life and will have to struggle to readjust ourselves to circumstances in order to regain it. The life of others in our community depends on this trust in ourselves and in them. People who prove themselves particularly trustworthy become its pillars, while those who wilfully transgress the rules and cannot be trusted drift to or are forced to its margins.

We find our place in the communal web by coordinating the requirements of this web and our innate or acquired talents, skills, and predilections. From our actual performance of our tasks we draw convictions about our dispositions, tastes, and character and come to believe ourselves to be decent, honest, sincere, etc.

In moments when we examine carefully our actions and conscience we recognize sincerely some shortcomings with respect to the established rules. We try to understand the reasons for these shortcomings and nearly always find some excuses which allow the basic conviction of our decency and honesty to remain intact: "I tried my very best," "Had I known better, I would have . . . ," "I could not help myself," etc.

We become so confident of our character that we attribute to ourselves a moral force that would allow us to handle situations far above those in our everyday routine. We may even contemplate hazardous undertakings believing that we would be on top of difficult situations. We may dream about doing great deeds that we would brilliantly accomplish. Above all, we feel confident that never would we waver in the fulfillment of our basic duties or fail others in the basic trust they put in us.

Hence, the moral pettiness which prevails in community opinion — and with which we have been dealing though this entire chapter. People avoid challenging this pettiness because they feel too weak to press their rights. They accomodate themselves to injustice and abuse, and instead of facing its significance directly they will even condone it. In order to spare their energies, nerves, and feelings and maintain peace of mind, they avoid confrontations; for the sake of peace they submit to the rule of the most cunning. Meanwhile, those who seek justice and truth and who challenge the rule of pettiness have to be prepared to fight difficult battles and to conduct innumerable intrigues against the cunning. In the end they might only fall into disrepute as troublemakers if they lose their case.

Those who for some reason find conformity difficult either remain at the fringes, where they are tolerated or even supported by the mainstream or wilfully stray into an "unruly" life. Responsibility, honesty, reliability, and other virtues seem to be commonly shared by all, and we naturally assume them in others as well as in ourselves. Rare are the occasions, though they will occur in a lifetime, when we are put on the spot. Then, and under some seemingly small provocation, we will break from the established code and system and will reveal quite unexpected features of our character, disposition, nature. Astounded, we face these — face ourselves — asking, "Who are we in truth?"

Indeed, we have to consider a most significant fact, namely, that the establishment of the individual within the societal web in accordance with its rules, regulations, principles, and laws of conduct involves the entire functioning of his/her living beingness. The entire set up of his faculties, psychic dispositions, feelings, sensitive reactions, and of the entire nervous and sensory-motor apparatus, of their functioning, is oriented by societal roles and situations; they subtend the weaving of its fabric. They attune our societal functioning; they bear it. Only through involving the complete spread of the various coordinated sectors of our functional circuits is the in-weaving process of our beingness accomplished, and this stable position of being a loop within a net or knit is crowned with our tacit, self-understood, safe, and self-confident state; this latter expresses, in fact, the harmonious "natural" cooperation that is being established.

With the breaking of one of the threads in a circuit, the weave, the entire system, becomes to some degree upset. The break calls for mending the situation, for the readjustment of other circuits, for reattuning, the reestablishment of harmony. Given the difficulties of accomplishing this, the individual might falter in his convictions about himself: he might fall out of society's web and drift to its margins in one or another way: moral or pathological, or a combination of both.

In onshore life, drastic situations and ensuing confrontations are mitigated by the circumstance of being enmeshed in a communal web that spreads far wider than the actual "encounter" between people. The vast network allows for prolonged negotiations, postponements, the manipulation of the particulars, counselling, the rearrangement of data, the cancelling out of interpretations, and mediating devices. In short, a man may go through life — and most do — without ever finding himself confronted with any situation in which in an unavoidable and irrevocable fashion he has to "face the facts" in such a way that his "true" standing on things is put to a test. But is there such a "true" standing?

With man being fashioned with reference to societal rules, where would he find other points of reference? In what could his very own "true" self be grounded if not in communal functioning? Can there be any "absolute" — if not standard — point of reference for one's very own truth?

As we have seen in the story of Jim, the sea is not in man's nature. On the sea man enjoys a merely provisional status, he not being an aquatic animal. He is absolutely dependent on his vessel and its condition. And this vessel even if it have the most advanced machinery is, as we have seen in the famous case of the *Titanic*, at the mercy of the sea.

On board his vessel, at the wheel, and powering his engines, man may feel secure. However, this security is confined to a transient moment during which the human being is entirely confined to a tiny space. Just beyond, he is altogether subject to the dominion of a powerful element. Here, there is no postponement, no negotiation; there is no making of pacts with the forces of nature. If a vessel fail a human being, he is hapless — at the mercy of a capricious, indifferent, and powerful element that engulfs all without a hearing.

(f) *Virtue as the Subliminal Thread Conducting the Inwardly Held Ideal into Overt Conduct*

Jim desparately asks Captain Marlowe, "What would you have done? . . . What could I have done?" He asks in vain. It is a rhetorical question. There is, indeed, no simple bridge between the subliminal self endowed with heightened moral consciousness, reflection, and imagination, on the one hand, and nervous physical reactions on the other; just as there is no direct relation between lofty romantic dreams and dispositions, on the one hand, and the hard reality of facts, on the other. That is, Jim cannot even in the deepest scrutiny of his beingness find an explanation for an elemental state of soul that would account for the discrepancy between his ideals and his act. Yet this search releases in him a most powerful passion: the passion to be true to oneself. His inner tragedy, which is his quest to find his true self, lies not at the bottom of his heart; the tragic knot ties up various spheres of his beingness and can be untied only by his conduct — if a second occasion for a courageous deed is given.

The urge to find a clue bridging the gap between his inner being and his overt action sets his entire being on fire. The fabulation of Jim's tale centers on the core of his inner self. Its rhythmic cadences follow Jim's struggle subsequent to the Inquiry; first, the self-searching quest, and second, Jim's wringing out of his innermost core conduct proportionate to the inner propensities of his moral self. The progress of this struggle spins the thread that would link his subliminal self with its endowed propensity toward an elevated moral life with his actual acts in times of trial when the nervous system and vitally informed reactions exercise their own contrary pulls.

As we know, Jim took a second chance with his life within a barbarian community in a far off land, one far enough away to be ignorant of the intrigues of the civilized world. He has there an extraordinary opportunity to make himself a vitally useful member of a human community. Amidst the violent struggles that constantly shake this society, Jim establishes himself as a cornerstone of honest rules. In the midst of constant danger be undertakes and carries out this task not out of romantic bravery or juvenile fearlessness, but by forging within himself a measure of courage from a sustained moral purpose. We fellow in rhythmic cadence the inner workings of Jim's wringing out from his innermost core the moral forces for his conduct while making it correspond to the propensities of his moral self: sincerity of purpose, generosity, dedication to the good of others, probity, and magnaminous honesty. It is within the stages of this progress that we find what are the steps to be taken to cover the long distance between the elevated moral ideal and one's actual acts in times of trial. We find, in fact, that in order "to be true to ourselves," that is, *true to our moral self in our conduct*, we have *to find a linkage between our subliminal sphere and the decision-making system of the will and to work out a model of conduct that will inform the acts that will spring forth*.

Then the passion to be true to oneself sets into motion the entire human functioning, orchestrating this complex process; it is also the passion to be true to oneself which functions as moderator and conductor in the accomplishment of this mobilizing of our entire being in which all our functions are shaped for the instantaneous attunement of our acts in a line of conduct that is faithful to our moral stand.

Have we not reached here the heart of our inquiry? What are the ideals and virtues we are talking about that great literature sets before us for our adherance? The concrete challenge to Jim, one of us, to be able to act "honorably," that is, to put into practice at every instant, if need be, the principles of the virtues which he believes in, has allowed us to bring forth the following insights: virtues such as courage, heroism, honesty, probity of judgment, selflessness — and even generosity, which is as we have shown the fountain of them all — are neither an inner principle (*arete*), as was believed in classical philosophies, nor a deeply rooted habit, disposition, or quality of our psyche. We have seen, in fact, that first the capacity to act according to our moral principles — a capacity which characterizes all virtues — consists in an orchestrated shaping of our entire functioning and that, second, "virtue," as the exercise of capacity, consists in *a dynamic thread uniting the subliminal moral ideal with the deliberative and prompting forces of the will and with those forces of our functioning that shape our conduct.*

To traverse these distances between the extreme of subliminal feelings, at the one pole, and psycho-physically incarnated action, at the other, that is, to practice any of the virtues, generosity included, we have to be prompted and conducted by the crucial passion of the innermost desire to be true to ourselves.

But we have not yet come to the end of our query. As we know, Jim has gained, through his steadfast reliability and responsible leadership, the absolute trust of the people; his self-confidence shows forth his self-reliance and feeling of fulfillment. Did this for him mean that he had redeemed himself before the human community? And yet to the admiring pronouncements of Marlowe who visits him in his far off land that he is a "better," a "truer" man than the rest, Jim bitterly snears, "but you would not have me on your ship."

In Jim's mind, the radical proof has not yet been obtained.

Did he not, however, obtain it with the final step in his quest when he went deliberately to submit himself to judgment, to a verdict that could only be death, out of sheer faithfulness, loyalty to his people, in order to keep his pledge to them?

(g) *The Subliminal Order of Virtues and the Continuity of Culture*

Our investigation into the inner tragedy of Jim, the sailor who "lost his honor" and set out to rehabilitate himself in the community of men, goes to the heart of the issues concerning culture.

First, what differentiates Jim from the others with which he committed his offence and what makes him be seen by people as being of a "better sort" is his level of culture. Moral consciousness, imagination, reflection, etc. differentiate him from the barbarians but also make him stand out in civilized society as an "uncommon man." Conrad shows us the difference between civilization and barbarism.

As Conrad presents it, primitive society is by no means an ideal state of natural relations guided by natural virtues. We have seen that, on the contrary, primitive man, incapable of protecting himself from the ruthless predations of unscrupulous men, struggles in vain to acquire some stability for himself and his family — for whose security he trembles. The rule of the stronger and the more cunning keeps ordinary people in constant fear. They submit to an arbitrary rule. The primitives spontaneously seek the stability of a reliable ruler, and they will entrust themselves unconditionally to the judgment of people who prove themselves either stronger, or more cunning, *or* more just. Yet with the development of culture, that is, with the development of the human capacity for reflection and the estimation of situations and with the organization of the societal order according to stable rules which grant the individual a role in the general pattern of society's functioning, the person does develop a core sense of personal responsibility and self-reliance along with an estimation and his/her own role and prerogatives.

We have given a summary sketch of the individual's interior involvement in and attunement to the societal order in the previous section of this inquiry — as a background against which we may disentangle the predicament of the "uncommon man" caught in a conflict with society and himself. Our description fits the society of our time, whereas Conrad's tale of "lost honor" would at first glance appear antiquated; honor, the virtues in general, do not seem to be the notions that now describe the principles according to which we model or judge our conduct.

The ideals and virtues of classical origin and those of the European culture which reached its peak in the literary expressions of the seventeenth century are no longer evoked. The terms which expressed them seem to have fallen into oblivion or at least to have lost their force in this "postmodern" period. Does this mean that the virtues and ideals for which they stand have vanished from the make-up of our persons? On the contrary.

To describe postmodern humanity at large in the terms of our study, there is no radical break between the high moral ideals of the great humanistic periods of Occidental culture and our times. We might today by and large ignore the high elevations of the spirit that were expressed in the ideals, virtues, and feelings of the classical and neoclassical periods of our literature/culture. The very notions of "ideals" and "virtues" might have lost their appeal and might seem to be replaced by more down to earth "values" of all sorts. Nevertheless, in describing the practice of the highly developed culture of Occidental postmodernity, we have enumerated some basic civic virtues that are actively assumed: dedication, responsibility, faithfulness, reliability, courage, sincerity, integrity, etc. These are all expected of the citizen. And in contrast to the classical period, these virtues are set forth as ideals not for people who are "larger than life" and who are of the cultural elite but for average ordinary people. The running of our societies is grounded in the realization of these common virtues by the great majority of people, and their concretization is considered an accomplished fact. Should we not consider them as variations of the ideals and virtues exalted in past cultural periods?

Conrad's doubt as to whether there is an "absolute" truth of the human self raises an ultimate question. I believe that we have answered his quest for an ultimate standard for the moral truth of the human being — truth which expresses the state of an inner self whose habits, tendencies, principles, and conduct are not relative to the rules and regulations of one or another established societal order but are uniquely his own. It is *in the unfolding of the subliminal propensities of the Human Condition* that we find the source of personal bents, strivings, modalities of feeling and conduct, virtues, and ideals. We find them there as they are carried by the deepest and most powerful of the subliminal passions of the human soul, the passion "to be true to oneself." These propensities carry in germ the principles of an inward moral realization which founds all the moral principles of personal and civic virtue and gives society its life and improves and advances its rules.

We may reject — and with the progress of culture, we do — various conceptions of the proper attitude toward life. In our postmodern period we hardly come across the concept of honor being employed as a criterion — standard — of conduct. It is, however, still sometimes evoked as belonging to the military code; "a soldier will not crawl" for rank or reward.

Nevertheless, the foundations of moral directives and criteria within the passional soul of the human being persist. They are always upheld by "the not so common man" — and they inform the criteria for self-esteem, self-respect, and dignity that underlie our conduct and win recognition in the community in which we live. Though sometimes the passion to be true to oneself degenerates into a soul-searching quest for one's "identity," or for "authenticity," that goes in futile directions, and though it is sometimes misled by wilfullness and lack of discernment, Occidental man's quest for "being true to himself" is more alive than ever. It manifests itself in his conduct in life as well as in his/her aversion to being willy-nilly directed by the judgments of others and in his will to exercise his own judgment using his own wits, upon his own evidence and within his life circumference. Albeit in a less poetic and sublime fashion, we meet in the midst of life heroic and virtuous men and women who are "one of us."

NOTES

1 Miguel de Unamuno, *Como se Hace una novela* (Madrid: Alianza Editorial, 1985), p. 120.
2 *Ibid.*, pp. 119—120.
3 *Ibid.*, p. 119.
4 Italo Calvino, *The Uses of Literature*, trans. Patrick Creagh (New York: Harcourt, Brace, Jovanovich, 1987), p. 39.
5 *Ibid.*, p. 41.
6 *Ibid.*, p. 44.
7 *Ibid.*, p. 45.
8 *Ibid.*, p. 49.
9 Ernesto Sabato, *Abaddón el exterminador* (Barcelona: Editorial Seix Barral, 1985).
10 Anna-Teresa Tymieniecka, *Logos and Life*, Book 1, *Creative Experience and the Critique of Reason* (Dordrecht, Boston, London: Kluwer Academic Publishers, 1988).
11 Georges de Scudery, "Observations sur *Le Cid*," April 1637.
12 Mathelot, *Memoires*, Champoin 1920.
13 August Strindberg, "Author's Foreword," *Miss Julie* in *Six Plays by Strindberg*, trans. E. Sprigge (Garden City, N.J.: Doubleday Anchor Books, 1955), p. 63.
14 *Ibid.*, p. 65.
15 *Idem.*
16 *Ibid.*, p. 66.
17 Herman Melville, *Moby Dick, or the White Whale* (Norwalk, Conn.: The Easton Press, 1977), p. 173.
18 *Ibid.*, p. 173.
19 *Ibid.*, p. 174.
20 *Ibid.*, pp. 85—86.
21 *Ibid.*, p. 178.
22 *Ibid.*, p. 557.

[23] *Ibid.*, p. 558.
[24] *Ibid.*, p. 581.
[25] *Ibid.*, p. 582.
[26] Joseph Conrad, *Lord Jim* (New York: The Crowell-Collier Publ. Co., 1962), p. 33.
[27] *Ibid.*, p. 82.
[28] *Ibid.*, p. 52.
[29] *Ibid.*, p. 89.
[30] *Ibid.*, p. 54.